Introduction to
ENVIRONMENTAL
REMOTE SENSING

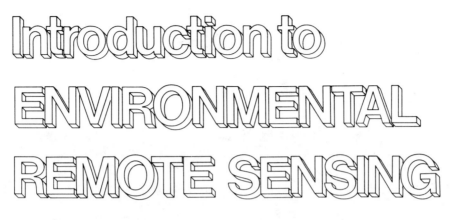

Introduction to ENVIRONMENTAL REMOTE SENSING

E. C. Barrett and L. F. Curtis

Department of Geography
University of Bristol

LONDON: CHAPMAN AND HALL
New York: John Wiley & Sons Inc.
A Halsted Press Book

First published 1976
by Chapman and Hall Ltd
11 New Fetter Lane

© 1976 E.C. Barrett and L.F. Curtis

Typeset by Josée Utteridge
of Red Lion Setters
and printed in Great Britain by
Fletcher and Son Ltd., Norwich

ISBN 0 412 12920 5

Distributed in the U.S.A.
by Halsted Press, a Division
of John Wiley & Sons, Inc., New York

Library of Congress Cataloging in Publication Data

Barrett, Eric Charles.
    Introduction to environmental remote sensing.

    Includes bibliographical references and index.
    1. Geography — Remote sensing. 2. Remote
sensing systems. I. Curtis, Leonard F., joint
author. II. Title.
G70.4.B37   1976   621.36'7   76-40968
ISBN 0-470-98959-9 (Halsted)

Contents

Preface

We believe this is a timely book. Remote sensing of the environment has experienced an explosive growth in the past two decades, but by nature it has been essentially a research science. Not surprisingly most early volumes emanating from this expansive field of enquiry have been edited collections of research papers or review articles. Very few have been integrated, general accounts for student textbook use. This new volume introduces the reader to the kinds of remote sensing techniques presently available for environmental analysis whilst also providing examples of their applications. Our aim has been to meet the needs of many students, administrators and scientists for an account of remote sensing which is essentially basic, explanatory and broad in its coverage. We have been forced to be selective and illustrative rather than exhaustive in our outlook. Nevertheless we have attempted to present students and practitioners of environmental science with an overview which might help them to seize opportunities for its application. As staff members of a Department of Geography we are glad to think that there are many examples of the practical applications of the discipline of Geography in the subject matter of this book. We are equally glad, however, to know that many other scientists find the powerful new tools of remote sensing to be both exciting and rewarding. Indeed remote sensing is nothing if it is not a multidisciplinary pursuit.

Our thanks are due to many friends and colleagues for sharing with us their own interests and professional conclusions. Some of these and others have kindly permitted their work to be summarized or illustrated here. Due acknowledgements are included in the text, in captions to illustrations and in end-of-chapter references. Our special thanks are due to the National Aeronautics and Space Administration of the U.S.A., the European Space Agency and the Department of Industry, U.K. for the ready availability of studies carried out in the fields of remote sensing.

Lastly, our thanks go to our wives Gillian and Diana for their continuous encouragement and support and their practical help with the manuscripts and proofs.

E.C. Barrett   L.F. Curtis

1 Monitoring the environment

1.1 The concept of environment

Few people today can be unaware of the existence of what is popularly called The Environment. Many communicators have discussed it, by radio, on television, and in print. The Environment has been drawn to the attention of the world community through recent gatherings like the United Nations Symposium on the Human Environment in Stockholm in 1972, and by international recognition of the first World Environment Day in 1974. At national, regional and very local levels increasing concern has been expressed about the need for improvement or protection of the better aspects of The Environment, and the urgency with which environmental exploitation or despoliation should be withstood. There is one problem, however, which is basic not only to such public discussion but also to the purpose and contents of this book: The Environment means different things to different people.

First of all we must recognize that there can be no such thing as an environment unless a situation is being studied from the point of view of the influence it has upon some selected object, either animate or inanimate, considered as an individual or a population. Certainly, for example, we may consider the environment of a single plant, or, indeed, at a still smaller scale, one of its constituent cells. From the point of view of the man in the street, The Environment is his own, or that of his fellow human beings, rather than that of another plant or animal. Alternatively The Environment can mean that of a town (often called its hinterland), or other objects considered in their spatial settings and in terms of various influences and interactions. It is with the environment of man, and especially with the resources it contains, that we shall be concerned in this book.

Sometimes man is discussed in the setting of the so-called 'natural' environment. This is frequently taken to be the same as his 'physical' environment, though, strictly speaking, it is more embracing in that it is concerned with other living things as well as the air, rocks and water which comprise man's purely physical home. Today it is often not enough to see our environment in natural terms. Man, with his large, powerful population and his great scientific and technological skills, has an enormous ability to exploit consciously, and modify both consciously and unconsciously (or accidentally) the world in which he lives. Without doubt he is the dominant life-form on his planet. In these days which are so marked by a growing awareness of the potential good – and the potential evil — we may bring upon ourselves through our interaction with our environment it is important to recognize that other people too are part and parcel of the smaller environments in which we live. We must not neglect to consider the 'cultural' environment if a better future is to be mapped out for the inhabitants of this rather overcrowded planet, which has been aptly likened to a little world island in a great sea of space.

Not surprisingly, it is often hard to dissect out those relationships linking man and his environment which are distinctly natural or cultural in type. Many chains of cause and effect are either intricately ramified, or inadequately known or understood. The great majority involve so-called feedback effects which often mean in practice that one man in trying to improve some aspect of his environment does so at the expense of some other aspect, or the environment of his neighbours. This kind of problem assumes enormous proportions in the economic sphere, through the discovery, evaluation, exploitation and use of natural resources — those constituents of the environment which are essential to the maintenance or improvement of our life-styles and standards of living.

In the compass of this book we shall be concerned with The Environment as if this were the environment of man, but acknowledging the role of man himself as both an influence upon it and a component in it. When assessing environmental matters, a number of stages may be followed in the investigation. These include:

(a) The recognition of the forms, structures, and/or processes of significance.
(b) The identification of such phenomena in their real world situation(s).
(c) The recording of their distribution, often through both space and time.
(d) The assessment of these distributions, sometimes singly, but more often in some combinations.
(e) Attempts to understand the nature and cause of any specially significant relationships.

Only when such background work has been successfully completed can the results be applied, commonly in one or more of the following ways:

(a) In the preparation and execution of schemes of environmental or resource management.
(b) In the planning and development of future projects.
(c) In the prediction of forthcoming events, over which one has less direct control.

At the heart of the all-important basic or background studies are problems concerned with observations. Data must be obtainable by appropriate means; they must be put into a permanent form; and there are great advantages if these data forms are such that they lend themselves ready to processing for analysis, comparison and interpretation. It is here that environmental remote sensing has such a vital part to play.

1.2 *In situ* sensing of the environment

Man has made measurements of key aspects of his environment since an early stage in the development of his civilisation and culture Examples of measuring devices include the famous Nilometer by which water levels in the River Nile were noted, and rudimentary rain gauges used by philosophers in the city states of Ancient Greece. The European Renaissance of art, science and literature marked a new surge of interest in the need for, and design of, monitoring instruments. Attention began to be paid not only to environmental variables which are readily visible or have immediately visible effects, but also to others less directly evident in the world of nature. The invention of the thermometer by Galileo at the end of the seventeeth century, and the mercury barometer by his pupil Torricelli at the beginning of the eighteenth century are specially notable.

A steady deepening of interest in environmental factors and conditions ensued through the eighteenth and nineteenth centuries before the rapid acceleration of the process in the twentieth century. Significant developments in the allied field of communications permitted the collection of data in near 'real time' (i.e. very close to the time of the observations) at a central processing facility. At last environmental studies could be organised on a broad scale, with the attendant possibilities of event prediction. More recently, the advance of other supporting technologies, especially in the field of electronic computers, has had a profound effect upon the design of *in situ* sensor networks and their capacities, and, through them, on the questions for which answers may be sought through analysis and interpretation of the resulting data.

1.3 Remote sensing of the environment

Remote sensing is the observation of a target by a device separated from it by some distance. In the case of *in situ* sensing, measuring devices are immersed in, or at least touch, the object of measurement. Some writers have alluded to instrument packages which are remote in the sense of being placed in relatively inaccessible regions, or have automatic data acquisition and transmission facilities, as remote sensing systems. This does not accord with the definition above, nor with the accepted view of remote sensing by those who practice it. Others have spoken of remote sensing in terms of a lack of physical contact between the sensor and its target. Again, strictly speaking, this is incorrect since, where no physical contact exists, no measurement is possible. While no 'touch' contact exists between a remote sensor and its target some physical emanation from, or effect of, the target must be found if aspects of its property and/or behaviour are to be investigated. The most important of the physical links between objects of measurement and remote sensing measuring devices involve electromagnetic energy, acoustic waves, and force fields associated with gravity and magnetism.

The invention of photography in 1826 may be said to have marked the opening of the remote sensing era, if the science of remote sensing is concerned as much with recording as with observation. As we shall see in Chapter 2 we all practice naturally some methods of remote sensing by which we observe the world in which we live. Devices like telescopes were invented long before the nineteenth century to extend our personal observing capabilities. But it is only in the last century and a half that means have been devised whereby the environment can be observed and recorded objectively by artificial devices. The potential of photography was quickly appreciated, especially for recording scenes out of reach of the observer. Later, new film types were invented which extended man's view beyond the narrow waveband of visible light. Then began a more systematic search through the radiation spectrum, and into the realms of acoustical, chemical, gravitational and radioactive energy to discover fresh means whereby normally insensible aspects of the natural and cultural environments might be investigated. This search was stimulated greatly by the military needs of two World Wars, especially the second, and has gathered even more momentum since.

It was in 1960 that reference was first made by name to remote sensing as a distinctive field of study, or set of approaches to the environment of man. Since then it has had all the appearances of the classical snowball, gathering momentum very rapidly, even collecting some accretions other than genuine snow. Much impetus was derived from the dawning of the satellite era and the space race between the remote sensing superpowers, the U.S.A. and the U.S.S.R.. In particular, the National Aeronautics and Space Administration (NASA) has played a large part through the ready availability of much of its data to the scientific community throughout the world. The Committee for Space Research (COSPAR) of the United Nations have also contributed significantly through their international conferences and working parties. More locally, the Michigan University Symposia on Remote Sensing of the Environment have become well established since the first in 1962, setting the pace for similar symposia in an ever-increasing number of centres around the world. The year 1969 saw the publication of the first specific journal, *Remote Sensing of Environment,* needed to take some of the growing strain off journals in germane sciences such as photogrammetry, geography, geology, geophysics, meteorology, and the biological sciences, to mention only a few. The early 1970's have witnessed the growth of new remote sensing societies in countries as far apart as the United Kingdom and Australia, and a mushrooming of courses in remote sensing in institutes of tertiary education. The extensive operations based on the first Earth Resources Technology Satellite (ERTS-1, renamed since as 'Landsat 1')* in 1972-4, and Skylab-A in 1973 have only confirmed an already widely-held belief that remote sensing has much to offer mankind in tending this rather small, worn planet for the common good.

* The ERTS program has recently been re-named 'Landsat'. We have retained the title ERTS-1 for the first Earth Resources Satellite since much of the extant literature refers to it as such. The title Landsat-2 will be used where the second satellite in this family is indicated.

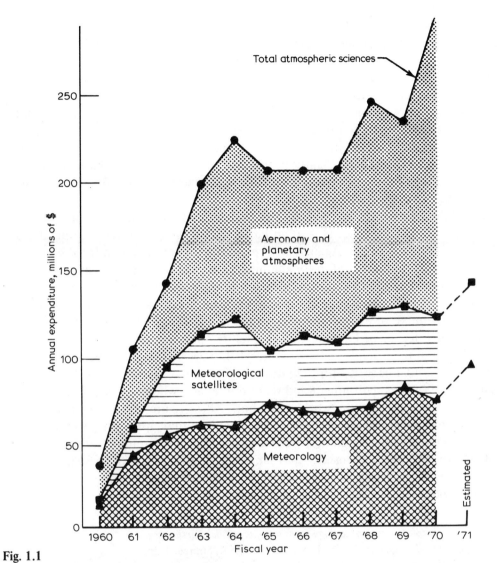

Fig. 1.1
American federal expenditure for atmospheric sciences research during the 1960's. (Source: Fleagle, 1971)

Although it is most difficult, especially in these days of rapid inflation and fluctuating currency values to place realistic figures on the costs and benefits which may be involved, some suggestions can be made which indicate roundly the kinds of sums involved. Fig. 1.1 indicates the order of magnitude of costs involved in the research aspects of the meteorological satellite system inaugurated in 1960 with the launching of Tiros I. It is in meteorology that the greatest use has been made of space data in day-to-day operations, and it is with relatively self-contained projects like Earth satellites that cost/benefit ratios can be most easily estimated. Three points are worthy of special note. First, the cost of the weather satellite programme undertaken by the Americans has stabilized since the mid-1960's at some $200m per annum. Second, the total annual cost to the American community of adverse weather is some $10 000m, of which 20 per cent represents a loss which could be eliminated given better weather information: this potential saving is greater than the current cost of weather satellite research by a factor of 10. Third, the improvements already made to hurricane warning systems average about $ 75m per annum — considerably more than the cost of the operational satellites — and may be much higher in individual years. Clearly in this area of environmental observation there are the strongest proven economic and humanitarian arguments for the remote sensing approach.

Estimates have been made by the U.S. National Academy of Sciences to assess the total cost — and the likely benefits — of an Earth resources satellite system operated in conjunction with meteorology. Against an expenditure of some £200m for the non-meteorological portion (and £180m for the meteorological segment) spread over a seven-year period, covering research and development, as well as initial hardware investment and operation, and maintenance in both the space and data distribution and usage sectors, a number of estimated savings may be set. Table 1.1 shows estimated savings in the first decade of Earth resources satellite operation over existing methods of gathering data in the disciplines listed. Table 1.2. suggests the orders of magnitudes of the potential annual benefits, through discovery of new resources and a more care-fully-planned use of the environment and its resources as a whole. They are almost certainly under-estimated for the world as a whole. These figures are a powerful argument for environmental remote sensing by satellites; they apply only to that part of the broader picture which may not be painted better by remote sensing from ground-based or low-altitude sensors. In other words, the total benefits which man may even now with proven technology derive from remote sensing of the environ-ment are vast indeed.

Table 1.1
Estimated annual savings from an operational Earth resources satellite system, 1975–1980. (After NASA, Source: Laing, 1971).

	Millions of £	
	U.S.A.	World
Agriculture and forestry	4 — 16	20 — 24
Geography	2 — 6	4 — 20
Geology and mineral resources	6 — 60	40 — 240
Hydrology and water resources	8 — 20	14 — 40
Oceanography	40 — 160	200 — 360
Totals	60 — 262	278 — 684

Table 1.2
Potential annual benefits from an operational Earth resources satellite system, 1975–1980. (After NASA, Source: Laing, 1971).

	U.S.A.	World
Agriculture and forestry	3×10^2	4×10^3
Geography	4×10^1	3×10^2
Geology and mineral resources	6×10^2	2×10^3
Hydrology and water resources	4×10^2	3×10^3
Oceanography	2×10^3	3×10^3
Totals	3.3×10^3	12.3×10^3

1.4 The geographical uses of remote sensing

Since the broadest and most co-ordinated use of remote sensing data has been made by the geographical community (see Fig. 1.2) it is of interest lastly to review the historic growth of remote sensing practices in geography, the most comprehensive of the sciences of man's environment.

(a) Pre—1925: a period of slowly increasing recognition of the potential utility of air photos in topographic mapping. The rise of aviation in the First World War accelerated the process to such an extent that already by the late 1920's some regions were being photographed systematically from the air.

(b) 1925—1945: a period of widespread but superficial use of aerial photography. During this span of twenty years air photo-interpretation became a fully-fledged intelligence-gathering technique for both civilian and military purposes. Many parts of the world were subjected to scrutiny from aloft, even hostile environments, relatively inaccessible on the ground, like much of Antarctica.

(c) 1945—1955: a period of preoccupation with interpretation techniques. This was a period during which many geographers in particular became enthusiastically acquainted with this new tool of great potential, and too much emphasis was placed on methods of analysis and interpretation, and too little on the applications to which the results could be put.

(d) 1955—1960: a period of widespread application of aerial photography. Now the theory began to be put to good use not only in topographic mapping, but also in various fields of human and physical geography, in geology, forestry, agriculture, archaeology, and numerous other disciplines with interests in spatial and temporal variation in the landscape.

(e) 1960—present: a period of active platform and sensor experiment. The first meteorological satellite in 1960 heralded the opening of a period of intense activity, investigating the potentialities of balloons, rockets, and especially

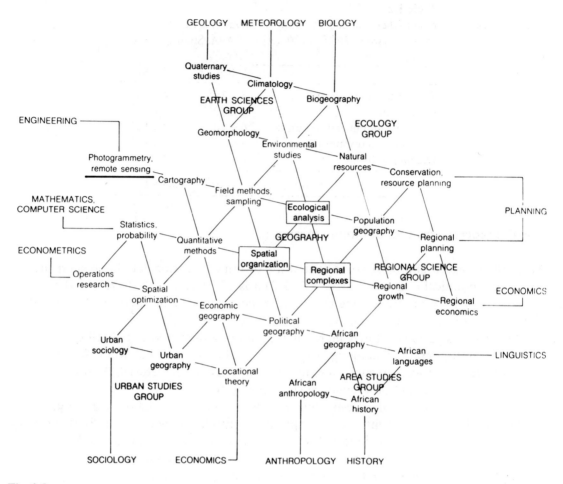

Fig. 1.2

Remote sensing as a gateway to the environmental sciences from the world of modern technology. In the cluster of Area Studies, Africa is chosen as an example only. (Source: Haggett, 1972)

satellites for remote sensing not only by conventional photography but also a wide variety of other means. ERTS-1 in particular (launched on July 23rd, 1972) yielded Earth surface data for many countries of the world in previously unrivalled quantities, and with unequalled homogeneity of quality.

Hopefully, the next decade will see remote sensing coming of age as a mature discipline, with a balanced development of theory, practices and applications. Now scientists, technologists, and politicians are so aware of the possibilities, the door has opened to planned, integrated and often international environmental studies based on fully appropriate instruments and sensor platforms, involving efficient processing techniques, and leading to valuable applications of the results.

References

Barrett, E.C. (1971), *Geography from Space,* Pergamon, Oxford.

Barrett, E.C. and Curtis, L.F. (eds), (1974), *Environmental Remote Sensing; applications and achievements,* Edward Arnold, London.

Estes, J.E. and Senger, L.W. (1974), *Remote Sensing; techniques for environmental analysis,* Hamilton Publishing Company, Santa Barbara.

Fishlock, D. (ed.), (1971), *A Guide to Earth Satellites,* Macdonald, London and Elsevier, New York.

Fleagle, R.G. (1971), 'The atmospheric sciences and man's needs: priorities for the future', *Bull. Amer. Met. Soc.,* **52**, 332.

Haggett, P. (1975), *Geography: a modern synthesis,* 2nd edn., Harper and Row, New York.

Laing, W. (1971), 'Earth resources satellites', in *A guide to Earth Satellites,* Fishlock, D. (ed.), Macdonald, London, and Elsevier, New York, pp. 69-91.

Lusignan, B. and Kiely, J. (eds.), (1970), *Global Weather Prediction: the coming revolution,* Holt, Rinehart and Winston, New York.

Stone, K. (1974), 'Developing geographical remote sensing', in *Remote Sensing; techniques for environmental analysis,* Estes, J.E. and Senger, L.W. (eds.), Hamilton Publishing Company, Santa Barbara, pp. 1-14.

2 Physical bases of remote sensing

2.1 Natural remote sensing

We all use our natural senses to observe and explore the environment in which we live. Certain senses — smell, taste, and, more often than not, touch — permit us to assess environmental qualities directly, through our neurophysical responses to the gases, liquids and solids with which we have immediate contact. The others — sight and hearing — can make us aware of more distant features through the patterns of energy propagations associated with them. Touch, manifested through the sensitivity of our skin to heat, also enables us to assess some characteristics of distant phenomena. These appreciations of the behaviour of energy sources some distance from ourselves are natural forms of remote sensing.

Since man's visual powers are, perhaps, his most highly prized means of gathering information about an object or phenomenon with which he is not in direct contact we may usefully examine these in more detail as an introduction to the principles involved in remote sensing by artificial means. The key components of the remote sensing system which makes sight possible are the eye and the brain. Visible energy, in the form of light emitted by, or reflected from, an illuminated object is detected by sensitive cells in the eye. The eye is linked by the optic nerve to a high speed, real time (i.e. near instantaneous) data processor — the brain.

The human eye is sensitive both to the intensity of the energy received, and the frequency of the wave-like perturbations which may be taken to characterize such flows of energy. As a consequence we can differentiate both ranges of brightness, and colour tone. In the brain the quickly-processed data are compressed and presented as a visual image. The brain also serves as a data bank in which earlier images can be stored, albeit within frustratingly narrow time limits, and with considerable loss of accuracy and definition. However, some qualitative pattern-matching can be carried out by drawing mental comparisons of, say, the present image and selected images recalled from the past.

Our mental pictures can be modified by simple means such as optical lenses to correct vision defects, selective filters like polaroid sunglasses to reduce the glare of strong sunlight, or by chemicals such as hallucinatory drugs to modify the way we perceive the area of illumination.

Clearly a natural remote sensing system such as the eye-brain is adapted best to the assessment of prevailing environmental patterns at the present time. In the evolutionary sense (where Darwinian factors operate) this is one of its greatest strengths. On the other hand, where conscious effort and freedom of choice are dominant this adaptation is one of its primary weaknesses. The chief requirements for artificial systems whereby man's remote sensing performance may be improved include:

(a) A broader and more selective ability to detect variations in environmental conditions. Our natural sensors have very limited performances, adjudged by what is physically possible, e.g. in the electromagnetic spectrum.

(b) A capacity for recording more permanently the patterns which are detected. This permits a more leisurely inspection of features of special interest.

(c) A better recall system so that patterns at different points in time might be compared with greater accuracy and in greater detail.

(d) The facility to play back the past at different speeds if required. Many natural events may be examined best when studied by time-lapse methods, or through a slow-speed play-back of recorded data.

(e) Opportunities for automatic ('objective') analyses of observations so that the personal (psychological, educational and/or neurophysical) peculiarities of the observer are minimized.

(f) Means of enhancing images to reveal or highlight selected phenomena.

Great strides have been made in recent years towards the fulfilment of such desiderata. Let us consider in more detail the range of opportunities presently being explored.

2.2 Technologically-assisted remote sensing

Over the years our limited natural capabilities for remote sensing have been much extended by the invention and steady improvement of a variety of specialized instruments. Here we shall confine our attention to the broad fields in which such instruments are designed to operate. Later in this chapter and elsewhere we shall go on to examine some of the instruments themselves, the forms and characteristics of their data, and ways in which these may be processed, analysed and interpreted.

2.2.1 Electromagnetic energy

Undoubtedly the most important medium for environmental remote sensing is electromagnetic radiation. This, the only form of energy transfer that can take place through free space, has enormous variety in its behaviour and implications. So important is this field that some authorities have considered it to be the only one. More than one account of remote sensing contains a categorical statement such as this: 'In remote sensing information transfer from an object to a sensor is accomplished by electromagnetic radiation'. Whilst other modes of information transfer certainly do occur (see below), we, too, shall concentrate our attention in this book on electromagnetism as a medium for remote sensing. This is justified by the preponderance of present interest in this field.

2.2.2 Acoustical energy

Acoustic sounding of the Earth's oceans is well-established as a means of profiling the sea floor. In a typical acoustic echo sounder short pulses of audible sound are emitted in a selected direction, and the returning reflected or scattered signals are collected as indications of ocean depth and sea floor characteristics. The atmosphere is also amenable to investigation using acoustic sounding techniques, especially aspects of its wind and thermal structure (Plate 2.1). Indeed, the strength of interaction of acoustic waves and the atmosphere is far greater than for electromagnetic waves, though the operational range of acoustic waves is less, and their slower speed of propagation can be a source of error with some measurements. Consequently the applications of such techniques to the atmosphere have been predominantly in research. For the future, the simplicity and much lower cost of an acoustic echo sounder system should ensure the continued development of such systems with eventual objectives in operational meteorology.

2.2.3 Force-fields

To the Earth scientist the most familiar force-fields are those of gravity and magnetism. Gravity is the force exerted by each body in the universe on every other body. It is directly proportional to the product of the masses of any pair of bodies, and inversely proportional to the square of the shortest distance between their centres of mass. We do not understand what causes gravity, but we can measure it, and use the results, for example to substantiate geophysical theories of the solid Earth and to reveal facts concerning the nature and disposition of its crust.

Magnetism is the attraction which certain natural minerals — and the Earth which contains them — have for others, especially iron. The magnetic properties of such minerals permit us to prospect for commercial deposits of them in geomagnetic surveys. Slow changes in the magnetic field of the Earth accompany the 'wanderings' of the magnetic pole; wilder variations are caused by fluctuations in the 'solar wind' of charged particles arriving from the Sun. These are responsible for 'magnetic storms' and the associated interference to radio communications, and the aurorae. They can be monitored by magnetometers to reveal and record changes in the behaviour of the Sun.

2.2.4 Active remote sensing

A distinction is commonly drawn between 'active' and 'passive' remote sensing. In the latter (illustrated earlier by the eye-brain system) the sensor detects the energy or force which emanates either from the target itself, or from a 'third party' source. Active remote sensing involves the detection of a signal which is artificially produced. Since such signals are generated under controlled conditions much can be learned from the ways in which they are affected by objects in the environment, and/or by the media through which they are transmitted. Radar (Radio Direction and Ranging) is a common example of an active system exploiting electromagnetic radiation. Sonar (Sound Navigation Ranging) is an active system utilizing acoustical energy.

2.3 Electromagnetic energy

2.3.1 The nature of radiation

Energy is the ability to do work. It can exist in a variety of forms, including chemical, electrical, heat, and mechanical energy. In the course of work being done, energy must be transferred from one body or one place to another. Such transfers are effected by:

(a) Conduction. This involves atomic or molecular collisions;

(b) Convection. This is a corpuscular mode of transfer in which bodies are themselves physically moved; and

(c) Radiation. This is the only form in which electromagnetic energy may be transmitted either through a medium or a vacuum.

It is this third type of transfer with which we are primarily concerned in remote sensing studies.

In keeping with many other areas in the environmental sciences, remote sensing employs models (simplified or idealized representations of reality) to describe complex phenomena, situations or interactions. In the case of electromagnetic radiation two models are necessary to describe and elucidate its most important characteristics:

(a) The wave model. This typifies radiation through regular oscillatory variations in the electric and magnetic fields surrounding a changed particle. Wave-like perturbations emanate from the source at the speed of light (3×10^{10} cm s^{-1}. They are generated by the oscillation of the particle itself. The two associated force-fields are mutually orthogonal, and both are perpendicular to the direction of advancement.

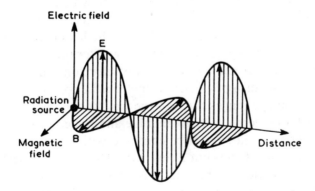

Fig. 2.1
Electric (E) and magnetic (B) vectors of an electromagnetic wave, viewed at a given instant. The intensity of radiation varies with the square of the peak amplitude of the electric field, and is proportional to the number of photons in the field. The type of radiation (see Fig. 2.2) is governed by wavelength, measured in absolute units.

(b) The particle model. This emphasizes that aspect of the behaviour of radiation which suggests it is comprised of many discrete units. These are called quanta or photons. These carry from the source some particle-like properties such as energy and momentum but differ from all other particles in having zero mass at rest. It has been hypothesized consequently that the photon is a kind of basic particle.

Whatever the true nature of electromagnetic radiation, we know that it is generated whenever an electrical charge is generated. In terms of the wave model, the *wavelength* of the resulting ray of energy is determined by the length of time that the charged particle is accelerated; the *frequency* of the radiation waves depends upon the number of accelerations per second to which the particle is subjected. The relationship between wavelength (λ), frequency (f), and the speed of light (a universal constant, c) is:

$$\lambda f = c, \text{ or } \lambda = c/f \tag{2.1}$$

This tells us that wave frequency is inversely proportional to wavelength, and directly proportional to its speed of wave advancement.

We may relate this expression of wave theory to the particle theory through the statement:

$$E = hf \tag{2.2}$$

where E is the energy of a quantum, h is a constant (named after M. Planck, who proposed the quantum theory in 1900), and f the frequency of the radiation waves. If we multiply ($\lambda = c/f$) by h/h (which does not alter its value), and substitute E for hf, we are left with a new expression, namely:

$$E = hc/\lambda. \tag{2.3}$$

This tells us that the energy of a photon varies directly with wave frequency and inversely with radiation wavelengths. This is borne out by experiment, for it can be demonstrated that the larger the wavelength of radiation the lower the energy involved, and conversely the higher the frequency of radiation, the greater the energy. These relationships are fundamental to the behaviour and appreciation of electromagnetic radiation.

Probably the most familiar form of electromagnetic radiation is visible light. Radiation detected by the human eye ranges through the well-known sequence of component colours from red through orange, yellow, green, blue and indigo to violet. This range, the so-called 'visible spectrum', should not be confused with the much broader 'electromagnetic spectrum', of which it forms but a tiny part. The eye-brain system is capable of detecting only a minute section of the total spectrum of radiation, all of which travels at the speed of light. Some knowledge of the breakdown of the electromagnetic spectrum is essential for an adequate appreciation of many remote sensing studies.

Largely for convenience of reference, the electromagnetic spectrum is subdivided into a number of sections. The boundaries between them are expressed in several different ways (see Fig. 2.2):

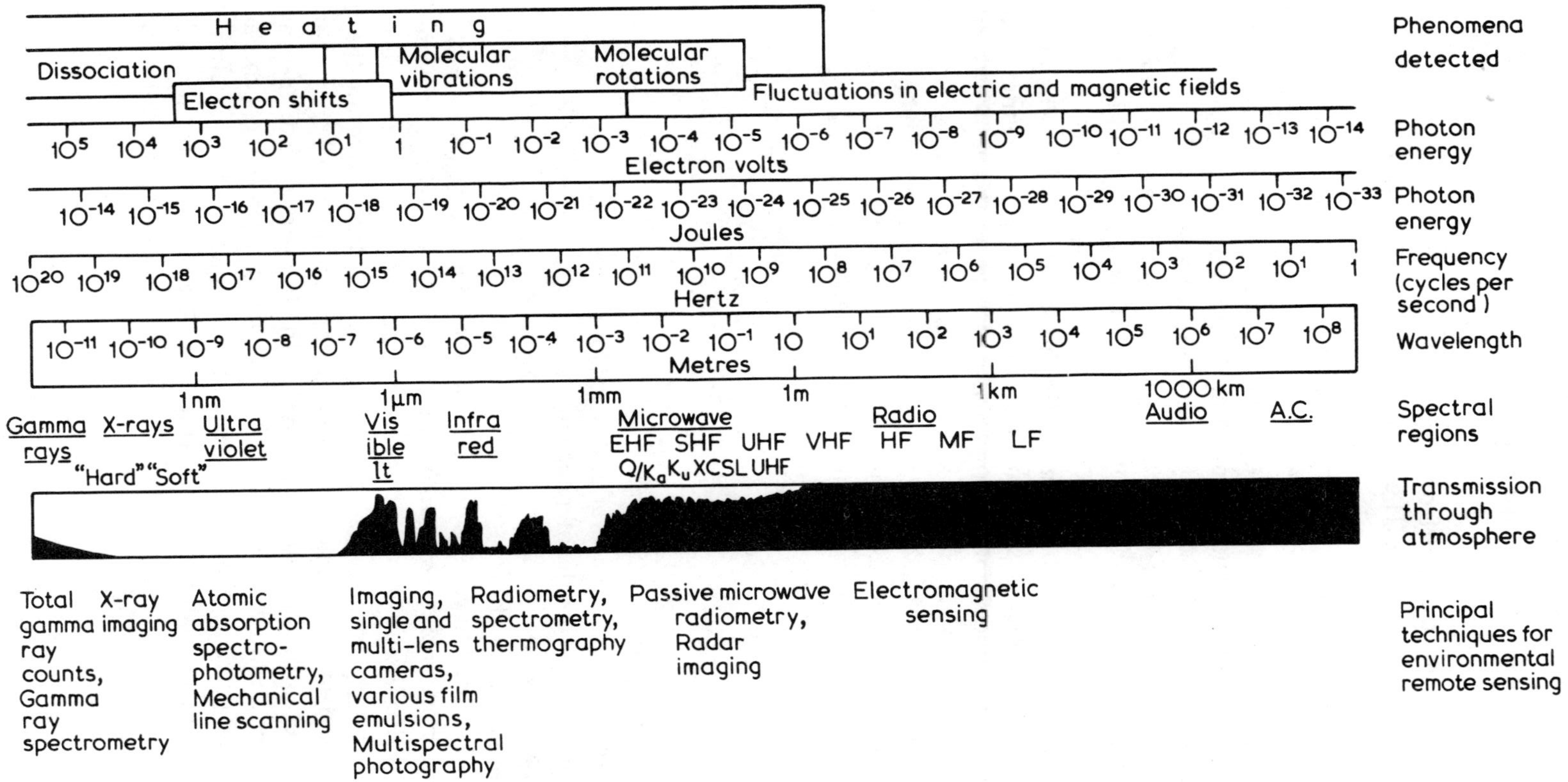

Phenomena detected

Heating — Molecular rotations — Molecular vibrations — Electron shifts — Dissociation — Fluctuations in electric and magnetic fields

Photon energy — Electron volts
10^{5} 10^{4} 10^{3} 10^{2} 10^{1} 1 10^{-1} 10^{-2} 10^{-3} 10^{-4} 10^{-5} 10^{-6} 10^{-7} 10^{-8} 10^{-9} 10^{-10} 10^{-11} 10^{-12} 10^{-13} 10^{-14}

Photon energy — Joules
10^{-14} 10^{-15} 10^{-16} 10^{-17} 10^{-18} 10^{-19} 10^{-20} 10^{-21} 10^{-22} 10^{-23} 10^{-24} 10^{-25} 10^{-26} 10^{-27} 10^{-28} 10^{-29} 10^{-30} 10^{-31} 10^{-32} 10^{-33}

Frequency (cycles per second) — Hertz
10^{20} 10^{19} 10^{18} 10^{17} 10^{16} 10^{15} 10^{14} 10^{13} 10^{12} 10^{11} 10^{10} 10^{9} 10^{8} 10^{7} 10^{6} 10^{5} 10^{4} 10^{3} 10^{2} 10^{1} 1

Wavelength — Metres
10^{-11} 10^{-10} 10^{-9} 10^{-8} 10^{-7} 10^{-6} 10^{-5} 10^{-4} 10^{-3} 10^{-2} 10^{-1} 10 10^{1} 10^{2} 10^{3} 10^{4} 10^{5} 10^{6} 10^{7} 10^{8}

1nm — 1μm — 1mm — 1m — 1km — 1000 km

Spectral regions
Gamma rays — X-rays "Hard" "Soft" — Ultra violet — Visible lt — Infra red — Microwave Q/K_a K_u X C S L UHF — Radio EHF SHF UHF VHF HF MF LF — Audio — A.C.

Transmission through atmosphere

Principal techniques for environmental remote sensing
Total gamma ray counts, Gamma ray spectrometry — X-ray imaging — Atomic absorption spectro-photometry, Mechanical line scanning — Imaging, single and multi-lens cameras, various film emulsions, Multispectral photography — Radiometry, spectrometry, thermography — Passive microwave radiometry, Radar imaging — Electromagnetic sensing

Fig. 2.2
The electromagnetic spectrum. The scales give the energy of the photons corresponding to radiation of different frequencies and wavelengths. The product of any wavelength and frequency is the speed of light. Phenomena detected at different wavelengths are noted, and the principle techniques for environmental remote sensing. The significance of the results for different environmental applications constitute a greater part of the latter chapters of this book: the student might usefully tabulate them for himself.

(a) Wavelength. The spectrum stretches from extremely short waves (cosmic rays) to very long ('electromagnetic waves'). The waves range in length from the microscopic to hundreds of kilometers.

(b) Frequency. We have seen that wavelength and frequency are always inversely related, since all waves advance at a common speed, the speed of light. Longwave radiation is a low frequency propagation, whereas short wave radiation is characterized by much higher frequencies. The units of measurement are hertz (Hz), or cycles per second.

(c) Photon energy. Equation 2.3 stated that the energy of a photon is directly proportional to wave frequency. Fig. 2.2 shows that long wave radiation has low photon energy, whereas shortwave radiation has high photon energy. This is expressed in joules or watt-seconds.

The whole spectrum ranges from almost infinitely short cosmic rays to the long waves of radio and beyond. It includes a number of familiar everyday manifestations such as gamma rays, X-rays, ultraviolet light, infrared rays and radar waves. Fig. 2.2 indicates some of their more common applications. It should be understood that the whole spectrum is a continuum which is subdivided in a rather inconsistent way. The boundaries we have between different regions of the electromagnetic spectrum are inexact. There is considerable overlap between some neighbouring regions, no fully-accepted terminology applies to them, and different authorities prefer different boundary values.

2.4 Radiation at source

2.4.1 The generation of radiation

An electromagnetic source whether natural or artificial may emit:
(a) A broad continuum of wavelengths of radiation;
(b) Radiation within a narrow (single spectral) band; or
(c) Radiation of a single wavelength.

The intensity of such radiation varies with the square of the peak amplitude of the electric field, and is proportional to the number of photons in the field. Variations in the intensity, and perhaps also the wavelength or frequency, of radiation from a source may change through time. The electrically-charged particles involved in generating the emitted radiation are basic units of matter, namely atoms, electrons and ions. All are in a state of constant motion under normal circumstances, when the temperatures of the objects they constitute are above absolute zero. Since the molecules of such objects contain those electrically-charged particles they possess natural resonance, either vibrational or rotational, so that they act as small oscillators which accelerate the electrical changes. The emitted photons of energy have frequencies which correspond to the resonances of the molecules in the meantime. The distribution of molecules among energy states is a function of temperature, so that at higher temperatures there are proportionately more molecules in the higher energy states. This is to say that all particles do not increase their energy equally as the temperature of a radiation source rises, but the overall effect is an increase in molecular activity, accompanied by an increase in emitted energy and a related shift in the dominant wavelength of radiation towards the higher frequencies.

2.4.2 Emission of radiation

All bodies with temperatures above absolute zero generate and send out, or emit, energy in radiant form. Each radiation source, or radiator (whether natural or artificial) emits a characteristic array of radiation waves assessed in terms of their wavelength and intensities. For many real-world radiators the array is very complex, comprised of a number of different contributions from the various constituents. Hence a characteristic curve or spectral signature may be obtained for each type of natural remote sensing target by plotting the intensities of the emitted radiation against appropriate wavelengths in the electromagnetic spectrum. Chapter 3 provides illustrations of a number of spectral signatures. For the present we must concern ourselves more with the principles which cause such signatures to differ.

A useful concept widely used by physicists in radiation studies is that of the blackbody, a model (perfect) absorber and radiator of electromagnetic radiation. A blackbody is conceived to be an object or substance which absorbs all the radiation incident upon it, and emits the maximum amount of radiation at all temperatures. Although there is no known substance in the natural world with such a performance, the blackbody concept is invaluable for the formulation of laws by comparison with which the behaviour of actual radiators may be assessed. These laws include the following:

(a) Stefan's (or Stefan-Boltzmann's) law. This states that the total emitted radiation (W, alternatively known as the radiant emittance) from a blackbody is proportional to the fourth power of its absolute temperature (T). We can express this as:

$$W = \sigma T^4 \tag{2.4}$$

When we speak of the 'total' radiant emission we are referring to the entire wavelength spectrum from $\lambda = 0$ to infinity. The constant (σ) relates to unit time and unit area. Commonly emittance is expressed in watts cm^{-2}. The thrust of Stefan-Boltzmann's law is that hot radiators emit more energy per unit area than cooler ones.

(b) Kirchhof's law. Since no real body is a perfect emitter, its emittance is less than that of a blackbody. Clearly it is often useful to know how the real emittance (W) of a radiator compares with that which would be anticipated from a corresponding perfect radiator (W_b). This may be established by evaluating the ratio W/W_b, which gives the emissivity (ϵ) of the real body. Thus, for the general case, we may say that:

$$W = \epsilon W_b \tag{2.5}$$

The emissivity of a real blackbody would be 1, whilst the emissivity of a body absorbing none of the radiation upon it (not unexpectedly known as a 'white body') would be 0. Between these two limiting values, the 'greyness' of real radiators can be assessed, frequently to two decimal places. If we plot the emittance curves for real radiators against corresponding blackbodies, we usually find that the idealized curves are much smoother and simpler than the observed patterns. In effect the emissivity index is a measure of radiating efficiency across the spectrum as a whole.

(c) Wien's (displacement) law. This states that the wavelength of peak radiant emittance (λ_{max}) of a blackbody is inversely proportional to its absolute temperature (T):

$$\lambda_{max} = C_3/T \tag{2.6}$$

C_3 is a constant equal to 2897 μm K. This equation tells us that, as a blackbody's temperature increases, so the dominant wavelength of emitted radiation shifts towards the short wavelength end of the spectrum. This can be exemplified by reference to the Sun and Earth as radiating bodies. For the Sun, with a mean surface temperature of about 6000K, $\lambda_{max} \simeq 0.5\ \mu$m. For the Earth, with a surface temperature of about 300K on a warm day, $\lambda_{max} \simeq 9.0\ \mu$m. Empirical observations have confirmed that the wavelength of maximum radiation from the Sun falls within the visible waveband of the electromagnetic spectrum whilst that from the Earth falls well within the infrared. We experience the former dominantly as light, and the latter as heat.

(d) Planck's law. This more complex statement describes accurately the spectral relationships between the temperature and radiative properties of a blackbody. In one form, Planck's law may be expressed by

$$W_\lambda = C_1\ \lambda^{-5}/(e^{C_2/\lambda T} - 1) \tag{2.7}$$

where W_λ is the energy emitted in unit time for our unit area within a unit range of wavelengths centred on λ, C_1 and C_2 are universal constants, and e the base of natural logarithms (2.718). A useful feature of Planck's law is that it enables us to assess the proportions of the total radiant emittance which fall between selected wavelengths. This can be useful in remote sensor design, and also in the interpretation of remote sensing observations.

In consequence of these four laws we see in Fig. 2.3 that emittance curves are characteristically negatively skewed, with their peak intensities falling in the lower quartiles of their spectral bands. The wavelengths of maximum emittance are progressively shorter for the hotter radiators, and the total emissions from the hotter radiators are greater than those from the cooler radiating bodies.

2.5 Radiation in propagation

2.5.1 Scattering

It is unfortunate (for remote sensing) that considerable complications arise in practice where the Earth's atmosphere intervenes between the sensor and its target. Although the speed of electromagnetic radiation is unaffected by the atmosphere, this medium may affect several of the other characteristics of this form of energy propagation. These include:

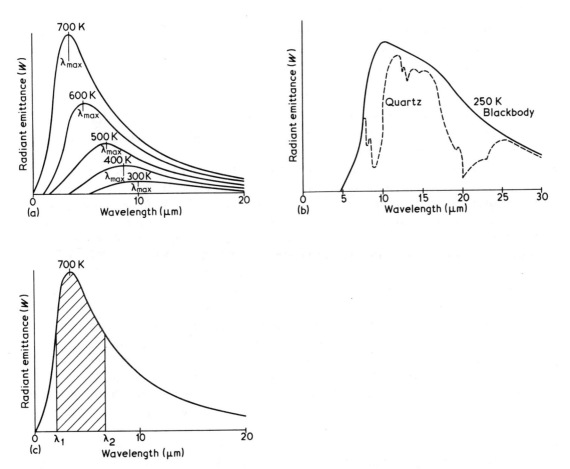

Fig. 2.3
(a) Selected black body radiation curves for various temperatures. Note that the areas under the curves diminish as temperature decreases. Note, too, that λ_{max}, the wavelength of peak radiant emittance, shifts towards the longer wavelengths with decreasing blackbody temperature.
(b) The total radiant emittance of a real body (e.g. quartz) is less than that of a blackbody at the same temperature. The general relationship is expressed by Kirchhof's Law.
(c) One form of Planck's Law of radiation permits the assessment of that portion of the total radiant emittance falling between selected wavelengths, e.g. λ_1 and λ_2

(a) The direction of radiation;
(b) The intensity of radiation;
(c) The wavelength and frequency of the radiation received by a target, at the base of the atmosphere;
(d) The spectral distribution of this radiant energy.

Frequently both the direction and intensity of radiation are altered by particles of matter borne in the atmosphere. These redirect in a rather unpredictable way the radiation en route through a turbid medium. An understanding of radiation scatter is necessary for the selection of sensors or filters where certain effects are required, and where image degradation due to atmospheric impurities must be avoided or at least reduced.

The attenuation which results from scattering by particles suspended in the atmosphere is related both to the wavelength of radiation, and the concentration and diameters of the particles, the optical density of the atmosphere (discussed under Refraction on p. 22) and its absorptivity. The common types of scatter are:

(a) Rayleigh scatter. This mostly involves molecules and other tiny particles with diameters much less than the radiation wavelength in question. It is characterized by an inverse fourth power dependence on wavelength. Hence, for example, ultraviolet radiation about one quarter the wavelength of red light, is scattered sixteen times as much. This helps to explain the dominance of orange and red at sunset when the sun is low in the sky: the shorter wavebands of visible light are cut out by a combination of atmospheric absorption and powerful scattering.

(b) Mie scatter. This occurs when the atmosphere contains essentially spherical particles whose diameters approximate to the wavelengths of radiation in question. Water vapour and particles of dust are the main agents which scatter visible light.

(c) Nonselective scatter. Here particles with diameters several times the radiation wavelengths are involved. Water droplets, for example, with diameters ranging commonly from 5–100 μm scatter all wavelengths of visible light (0.4–0.7 μm) with equal efficiency. As a consequence clouds and fog appear whitish, for a mixture of all colours in approximately equal quantities produces white light.

2.5.2 Absorption

This is the process by which radiant energy is retained by a substance or a body. In the real world it involves the transformation of some of the incident radiation into heat, and the subsequent re-emission of that energy at a longer wavelength.

Considering the ideal case, it will be recalled that a blackbody is both a perfect radiator and fully efficient absorber. As the emissivity of a real body can be expressed as a ratio of the (ideal) blackbody performance, so also can the absorptance (a) be expressed. Hence, for a blackbody, both ϵ and a equal unity. In this ideal case $a_b = \epsilon_b$. The same relationship can be shown to hold good for 'grey' bodies if they are opaque.

In the atmosphere — which can be decidedly murky, but is never opaque — radiation absorption takes place not at its surface but in transit. Three gases, water vapour, carbon dioxide, and ozone, are particularly efficient absorbers of radiation from the Sun. Consequently incoming solar radiation (often abbreviated to 'insolation') — which is by far the most important natural source of radiation for passive remote sensing — is attenuated significantly by its passage through the atmosphere. Often the combined effects of absorption and scattering by particulate matter are expressed in terms of an 'extinction coefficient'. This enables us to evaluate the energy which reaches the surface of the planet

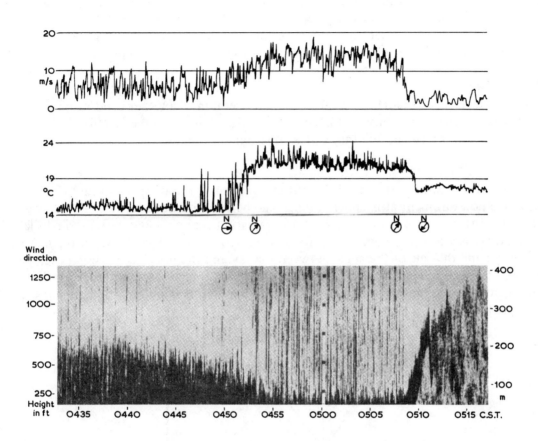

Plate 2.1

It is possible to detect the small amount of energy backscattered from a vertically directed beam of sound by inhomogeneities in the density structure of the lower troposphere. Plate 2.1 exemplifies the type of visual record obtained, as a function of height and time. This record was obtained during the passage of a cold front. The traces show wind (at top) and temperature (bottom). (From McAllister and Pollard, 1969)

Earth as a proportion of the radiation incident upon the outer limits of the atmosphere. This proportion permits us to evaluate the transmittance of the atmosphere. For any such medium transmission is inversely related to the extinction coefficient times the thickness of the layer. In practice, therefore, the transmittance decreases as the combined effects of absorption and scattering increase, perhaps as the transit medium increases in depth.

2.5.3 Refraction

When electromagnetic radiation passes from one medium to another bending or 'refraction' occurs in response to the contrasting densities of the media (Fig. 2.4). A measure of this is given by the Index of Refraction (n), where:

$$n = c/c_n \tag{2.8}$$

The index is the ratio of the speed of light in a vacuum (c) to its speed in the substance (c_n). In a non-turbulent atmosphere (which can be conceived as a series of layers of gases each of a different density) the refraction of radiation is predictable. For, as Snell's law states, there is a constant relationship for a given frequency of light between the index of refraction and the sine of the angle between the ray and the interface between each pair of adjacent layers. Problems do arise, however, where a turbulent atmosphere is involved. Turbulent motions are essentially random, and their effects upon radiation are consequently unpredictable.

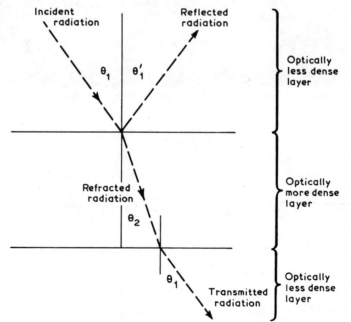

Fig. 2.4
The relationships between incident radiation, reflected radiation, refracted radiation and transmitted radiation, and a stratification of optically different layers.

2.6 Radiation at its target

2.6.1 Reflection

So far we have discussed the effect of particles on radiation as if the redirection is again something we cannot predict. In the real world this is not always so. The critical factors are the smoothness and orientation of the object lying in the path of an electromagnetic ray. If the object has smaller surface irregularities than the wavelength of the impinging energy, it will act as a mirror and the angle at which the energy is directed away from the object will equal the angle of its incidence upon it. This process of reflection is sometimes called backscattering. The angles of incidence and reflection, and a perpendicular to the reflecting surface from which those angles are measured all lie in the same plane. Conversely scattering may be described in terms of reflection, when it is labelled diffuse reflection in contrast to specular reflection which is the ideal case. Although we might expect

Table 2.1
Values of albedoes for various surfaces and types of surfaces.
(Source: Lockwood, 1974).

Type of surface	Surface	Albedo % of incident shortwave radiation)
Soils	Fine sand	37
	Dry, black soil	14
	Moist ploughed field	14
	Moist black soil	8
Water surfaces	Dense, clean and dry snow	86–95
	Woody farm, snow-covered	33–40
	Sea ice	36
	Ice sheet with water covering	26
Vegetation	Desert shrubland	20–29
	Winter wheat	16–23
	Oaks	18
	Deciduous forest	17
	Pine forest	14
	Prairie	12–13
	Swamp	10–14
	Heather	10
Geographic locations	Yuma, Arizona	20
	Winnipeg (July)	13–16
	Washington, D.C. (September)	12–13
	Great Salt Lake, Utah	3

some reflection by particles in the atmosphere, reflection is much more important for remote sensing where continuous (e.g. land or sea) surfaces are involved.

A useful expression of the reflectivity of different terrestrial surfaces is the reflection coefficient, or albedo (Table 2.1). The albedo of such a surface is the percentage of the insolation incident upon it which is reflected back towards space.

We referred earlier to a white body as one which absorbs none of the radiation which impinges on it. We are now able to say that, whereas a blackbody is conceived as a perfect absorber of radiation, a white body is a perfect reflector. Since none of the incident radiation is absorbed, the surface temperature of a white body remains unchanged.

Because visible spectrum wavelengths are so short, most surfaces reflect light diffusely regardless of the angles at which this radiation strikes them. Some of the reflected energy returns in the direction of its source. Longer wavelengths, e.g. microwaves, create more specular reflection off the same surfaces, the reflected energy being directed generally away from its source. Some of the consequences of such differences in performance may be illustrated by reference to the remote sensing of a building at night. Using visible light — a torch or a searchlight — the facing wall of the building will be discernible whatever its angle to the light beam. Using radar to locate the same feature much depends upon the angle between the radar beam and the wall it strikes. Since most of the reflection in this case is specular rather than diffuse, much stronger returns are obtained from a head-on, rather than an oblique, orientation.

Reflected radiation is of considerable importance to remote sensing since so many observing systems are based upon it (Plate 2.2.). The natural remote sensing system of the eye and the brain observes and perceives many aspects of the natural environment through reflection of the solar radiation by which it is illuminated. The early, now well-developed, artificial technique of photography records phenomena through the light which is reflected from them, whether in the studio or out of doors. Active remote sensing systems such as radar often measure the reflection of specially-propagated energy so that the sizes, positions and/or densities of natural or man-made reflectors can be assessed thereby.

2.6.2 Absorption

Not all the radiation incident upon the surface of a target is necessarily scattered or reflected by it. Some of the incident energy may enter the target to be propagated through it as a refracted wave front. As we noted earlier, with a gaseous medium like the atmosphere, energy in transit may be attenuated or depleted by the process of absorption. The ability of a substance to absorb radiation (i.e. its absorptance) depends upon its composition and its thickness.

Absorptance also varies with wavelength of radiation. A target may behave very differently if exposed to radiation of different wavelengths. For example, an object or a medium may be highly absorptive in the visible range yet transparent in the infrared (like some 'semi-conductors'), or it might be transparent in the visible and opaque in the infrared (like glass).

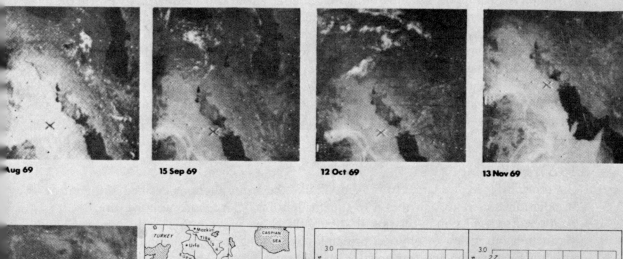

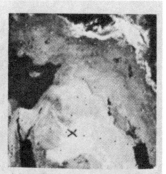
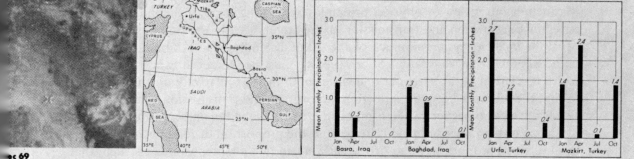
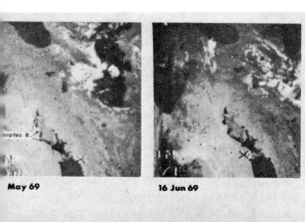

Plate 2.2

Nimbus III daytime HRIR pictures showing the effects of the summer minimum of precipitation on the levels and patterns of the 0.7–1.3 μm band reflected solar radiation over the Middle East. In the image for May 30th the Euphrates valley is still dark and wide with vegetation and soil moisture remaining from winter precipitation. As the summer progressed and precipitation decreased the river areas assumed an appearance like the surrounding steppes and deserts. By August 6th, the Euphrates in Iraq has become almost indistinct. The graphs show rainfall for Basra and Baghdad, along the river in Iraq, and for Urfa and Mazkirt in the highlands of Turkey. (Courtesy, NASA)

Clearly some knowledge of the absorptance of a target is often invaluable in choosing the best remote sensing means to solve a specific problem, and in interpreting correctly the results of a remote sensing survey.

The net effect of absorption of radiation by most substances — including the atmosphere and terrestrial surfaces — is that the greater part of the absorbed energy is converted into heat. Thereby the temperatures of the substances are raised. This heat energy may enable the original target of radiation to become a secondary radiation source itself, through its emitting some of the energy absorbed. Since the peak intensity of solar radiation is in the waveband of visible light, and the Earth/atmosphere system is by no means a blackbody, it is not surprising that its radiation temperature — as we saw earlier — is much less that of the Sun. The radiation peak of the Earth is in the infrared (see Fig. 3.1). Fortunately for practices in environmental remote sensing infrared energy (a form of 'thermal' radiation) is emitted from the surface of the Earth by both day and night, and is affected little by atmospheric particles such as haze and smoke. If we take care to view remote sensing targets through atmospheric window wavebands in the infrared portion of the spectrum (where absorption by certain gaseous constituents of the atmosphere is negligible) and under cloud-free conditions, we are able to add much to our knowledge of their physical and chemical properties which we could not have learnt by conventional photography. As we shall see in later chapters, thermal mapping by infrared and microwave techniques is becoming increasingly significant in remote sensing applications today.

2.6.3 Transmission
The transmittance of a target (or a medium like the atmosphere) is defined as the ratio of radiation at distance x within it to the incident radiation. Since we have already mentioned transmission in relation to other effects, we may, in this section, conveniently conclude our review of physical principles by revising some of the interplay between reflection, absorption and transmission at the target. Clearly the sum of the three must equal the incident radiation, much of the detail depending upon the nature of the target itself, whether it be transparent or opaque. We must not forget, however, that a target which is relatively transparent at one wavelength may be relatively opaque at another, so that the relations between reflection, absorption and transmission generally vary across the electromagnetic spectrum. Lastly, it should be remembered that the angle of incidence of the radiation may, if low, cause the proportion of reflected energy to exceed the combined proportion of absorbed and transmitted energy, whereas, if high, that angle may allow more of the energy to enter the target providing its detailed surface characteristics permit (see Fig. 2.4.).

2.7 General conclusions

Although it may have seemed at the beginning of this chapter that the earth scientist wishing to employ remote sensing techniques to improve his perception and appreciation of the human environment need do no more than choose the most appropriate instrument and the best platform for his purposes, it must be clear by now that many factors may influence that choice. In later chapters we shall review some of the more common sensors, sensor packages, and remote sensing platforms. However, it is clear that if we are to understand this rapidly expanding field of scientific activity, rather than just know of it, we must try to relate experimental and operational practice to the principles we have examined. Since this might be difficult without further explanation and illustration Chapter 3 is concerned with the observed radiation characteristics of some natural phenomena. This will provide a bridge between the rather abstract set of theoretical statements in Chapter 2 and the very technical array of practical possibilities discussed later in this book.

References

Feinberg, G. (1968), 'Light', *Scient. Am.,* **219**, 50.

Holz, R.K. (ed.), (1973), *The Surveillant Science; remote sensing of the environment, Part 1, The Electromagnetic Spectrum – energy for information transfer,* Houghton Mifflin, Boston, pp. 1-27.

McAllister, L.G. and Pollard, J.R. (1969), 'Acoustic sounding of the lower atmosphere', *Proceedings of the Sixth International Symposium on Remote Sensing of Environment,* Ann Arbor, Michigan, pp. 436-50.

Simon, I. (1966), *Infrared Radiation,* Van Nostrand, New York.

Weisskopf, V. (1968), 'How light interacts with matter', *Scient. Am.,* **219**, 60.

3 Radiation characteristics of natural phenomena

3.1 Radiation from the Sun

3.1.1 The spectrum of solar radiation

Solar radiation reaches the outer limit of the Earth's atmosphere undepleted except for the effect of distance. Since the total radiation from a spherical source such as the Sun passes through successively larger spheres as it radiates outward, the amount passing through a unit area is inversely proportional to the square of the distance between the area and the radiation source. Although the plant Earth, 93 million miles from the Sun, intercepts less than one fifty millionth of the Sun's total energy output this intercepted energy is vital not only to most remote sensing techniques — either directly or indirectly — but is vital also to terrestrial life itself.

Observations of the radiation output from the Sun are made difficult by the attenuating effects of the Earth's atmosphere, at or near whose base most measurements have been made. Clearly the best vantage point for evaluating the spectrum of solar radiation is *outside* the atmosphere. Although some data from satellite-borne sensors are now available probably the most reliable statement which is reasonably complete is still that obtained many years ago by staff members of the Smithsonian Institute from the elevated observatory at Mount Wilson. Rocket measurements of ultraviolet radiation can be added to extend the spectrum towards the short wavelength end, where solar radiation is almost completely blocked by absorption in the upper atmosphere. Fig. 3.1 summarizes the results, including a curve extrapolated to the outer limits of the atmosphere, and corrected for mean solar distance.

3.1.2 The atmospheric absorption spectrum

We remarked in Chapter 2 and again earlier in this chapter, that solar radiation is significantly attenuated by the envelope of gases which surrounds the Earth. Consequently the curve of solar radiation incident on the Earth's surface stands below that for the top of the atmosphere in Fig. 3.1. Some of the attenuated radiation is quickly lost to space, by scattering processes, whilst the remainder of it is absorbed by the atmosphere itself. Almost all the solar radiation which reaches the surface of the Earth is of wavelengths shorter than 4.0 μm, whereas that emitted by the Earth (see Fig. 3.1) is principally in the broad waveband from 4.0–40 μm. There is practically no overlap between these two types of radiation, which are sometimes referred to as shortwave and longwave radiation respectively. The atmosphere has the ability to absorb not only

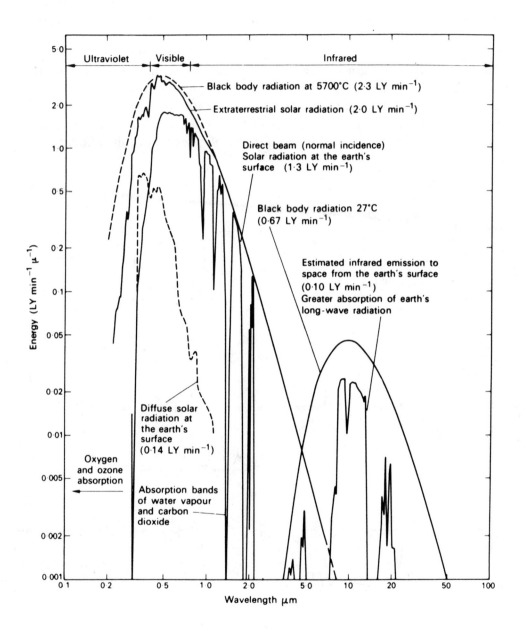

Fig. 3.1
Electromagnetic spectra of solar and terrestrial radiation. The curve for extra-terrestrial solar radiation represents that radiation which is incident on the top of the Earth's atmosphere for the mean distance between the Earth and the Sun. (Source: Sellers, 1965)

direct solar radiation, but longwave (terrestrial) radiation also. In considering the atmospheric absorption spectrum it is more convenient to review the impact this has upon solar and terrestrial radiation together rather than separately.

The patterns of absorptivity of different constituents of the atmosphere across part of the electromagnetic spectrum are portrayed in Fig. 3.2. The most striking feature of the set of curves as a whole is the great variability from one wavelength to another. In many cases sharp peaks of absorptivity at some wavelengths are separated by equally abrupt troughs where the ability of the gas to absorb radiation is very low. The most important absorbers of radiation within the mixture of gases we call the atmosphere are three in number:

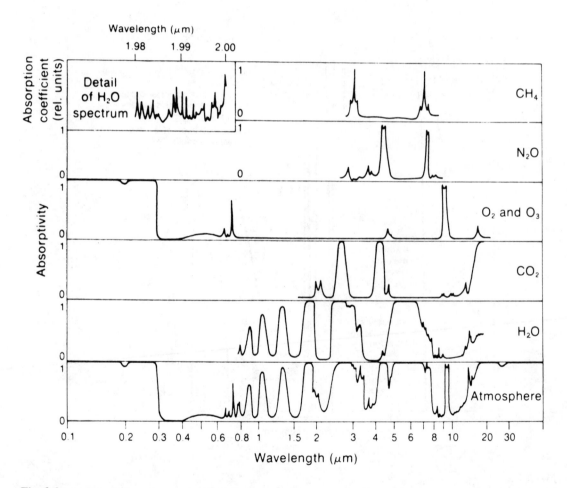

Fig. 3.2

Spectra of absorptivity by constituents of the atmosphere, and the atmosphere as a whole. (Source: Fleagle and Businger, 1963)

(a) Oxygen and ozone. Radiation of wavelengths less than about 0.3 μm is not observed at the ground. Almost all of that reaching the top of the atmosphere is absorbed high in the atmosphere: the rest is backscattered to space. Energy of $\lambda < 0.1$ μm is highly absorbed by O and O_2 (also by N_2) in the ionosphere; energy of 0.1–0.3 μm is absorbed efficiently by O_3 in the ozonosphere. Further, but less complete, ozone absorption occurs in the 0.32–0.36 μm region, and at a minor level around 0.6 μm (in the visible), and 4.75 μm, 9.6 μm and 14.1 μm (in the infrared).

(b) Carbon dioxide. This is of chief significance in the lower stratosphere. It is more evenly distributed than the other absorbing gases and it is overshadowed at high altitudes by oxygen and ozone, and in the troposphere by water vapour. It has weak absorption bands at about 4 μm and 10 μm, and a very strong absorption band around 15 μm which is well known and widely exploited for atmospheric sounding.

(c) Water vapour. Among the atmospheric gases this absorbs the largest amount of solar energy. Several weak absorption lines occur below 0.7 μm, while important broad bands of varying intensity have been identified between 0.7–8.0 μm. The strongest water vapour absorption is around 6 μm, where approaching 100 per cent long wave radiation may be absorbed if the atmosphere is sufficiently moist.

In Fig. 3.2, below the curves for individual constituents of the atmosphere we see the total atmospheric absorption spectrum. At the short end of this spectrum the atmosphere effectively absorbs almost all the incident radiation. However, it is largely transparent in the visible band from 0.3–0.7 μm. Thereafter in the direction of increasing wavelengths we see a sequence of more or less sharply-defined absorption bands alternating with relatively transparent regions. These transparent regions are the most useful windows in the absorption spectrum.

Broadly speaking, practitioners of environmental remote sensing whose major interests are in Earth surface features avoid those wavebands in which atmospheric absorption is strongly marked, especially if they plan to use high-level observation platforms. The wavebands of maximum absorption may, however, be chosen deliberately for some kinds of meterological and climatological studies. For example sounding of the atmosphere in depth by spectroscopic means has exploited the 15 μm CO_2-absorption waveband. In studies of this kind the aim is to identify and measure the radiation emitted from a number of levels in the atmosphere, so that vertical profiles of the structure of the atmosphere may be obtained. (See Chapter 9).

The basic principle on which satellite spectroscopy is founded is that the molecules of a gas emit electromagnetic radiation at the frequencies at which energy is absorbed. Absorption by a gas like CO_2 is due to many different vibrational modes of the gas molecules. These modes are arranged in a series of bands across the absorption spectrum. Spectroscopic sounding from a satellite or other high-level platform involves the measurement of radiation emitted by the underlying atmosphere at a series of frequencies where the atmosphere absorbs in a known way. Fig. 3.3. illustrates this approach by reference to a single frequency emitted vertically upwards. The frequency is chosen to give a 50 per cent transmittance from level B to the satellite, i.e. radiation packet (b) reaches the satellite half attenuated. Packet (a) is attenuated more on account of its passage through a greater depth of the atmosphere. Packet (c) arises at the satellite

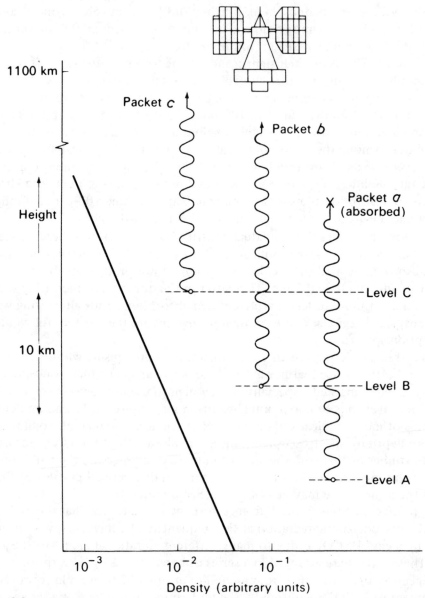

Fig. 3.3
The behaviour of packets of radiation emitted by carbon dioxide from different levels of the atmosphere. (Source: Barnett and Walshaw, 1974)

virtually unattenuated, but emission from this level is small: the emission of radiation by CO_2 is proportional to the concentration of the gas, which decreases exponentially with height. Hence the radiation measured by the satellite-borne sensor is a weighted mean of that emitted from various layers, with the greatest contribution from around level B. If the emitted radiation is measured at several frequencies, suitable weighting functions can be calculated to specify the atmospheric pressure levels from which most of the measured radiation originates (Fig. 3.4).

Spectroscopic soundings of the atmosphere have been made successfully by several Nimbus-borne sensors, e.g. the Satellite Infrared Spectrometer (SIRS) which has already provided many operationally-useful temperature profiles of the troposphere for short-term forecasting (see Chapter 9). Another successful sounding device is the Selective Chopper Radiometer (SCR) which has added much to our knowledge and understanding of temperature patterns in the stratosphere and above. Other sensor systems e.g. the

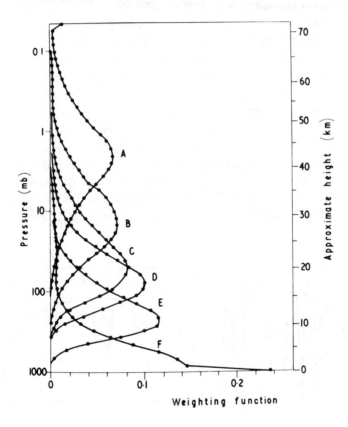

Fig. 3.4
The weighting functions of the Selective Chopper Radiometer (SCR) on Nimbus IV. The height scale is approximate, since the heights of the pressure surfaces vary by up to a few km according to the temperature profile. (Source: Barnett and Walshaw, 1974)

Infrared Interferometer Spectrometer (IRIS) have been employed not only to investigate the Earth's atmosphere, but also the atmospheres of other planets, for example Mars. Still further spectrometers have been tested or proposed, and, as with many other remote sensing systems designed initially for investigating our terrestrial home, there must be a bright future for at least some of them for very long range remote sensing to observe other units in the solar system. We already have a firm basis for studying in considerable detail the dynamics of atmospheres other than our own.

3.1.3 Atmospheric transmission

We need say little under this heading in view of the almost perfect inverse relationship between the atmospheric absorption spectrum and the spectrum of atmospheric transmission, other things being equal. A typical atmospheric transmission curve is shown in Fig. 3.5. Its upside-down similarity to Fig. 3.2 (allowing for the different horizontal scales) is very striking. It should be noted, however, that the effective transmittance of any atmospheric column is very dependent upon both absorption and scattering. The situation we have chosen to illustrate is one in which humidity is low, and there is little particulate matter in the column. Under hazy or foggy conditions attenuation by scattering may have a larger effect upon the spectral transmittance of the atmosphere at sea-level than attenuation by absorption. Consequently if one wishes to view the surface of the Earth from aloft, especially at the short end of the spectrum, care must be taken to choose those atmospheric conditions under which attenuation will be as little as possible. Then, transmittance will be at its peak. Often the angle of view will be critical for the decision whether or not to undertake any specific photographic mission. Under given attenuation conditions transmission is best for vertical paths, deteriorating through slant paths to the horizontal, where the view at the surface of the Earth is entirely within the layer characterized by the highest concentrations of atmospheric water and suspended particles.

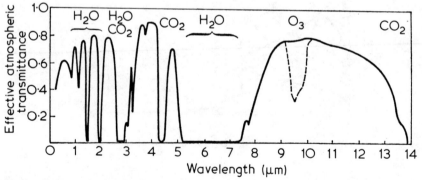

Fig. 3.5

The spectral transmittance of 2000 yards of sea-level atmosphere at low humidity and low haziness. The principal regions of absorption by certain gaseous constituents of the atmosphere are labelled.

3.2 Radiation from the Earth

3.2.1 Atmospheric window wavebands

Much of the planning of airborne or spacecraft missions in remote sensing of the planet Earth as distinct from its atmospheric envelope involves the identification and exploitation of the windows between bands of peak attenuation. We may illustrate the principles involved by reference to the two most useful windows in the infrared, extending from about 3.0–4.5 μm and from 8.5–14 μm. These can be identified easily in Figs. 3.2 and 3.5. The chief advantage of the first is that it is the more sharply defined. The chief advantage of the second is that it lies solely within the thermal emission spectrum of the Earth, and embraces the wavelength of peak infrared emission from that source (approximately 10 μm). Unfortunately there is a marked atmospheric absorption band around 9.6 μm but this is usually disregarded in all but satellite and high-altitude rocket missions. As Fig. 3.5 reveals, this absorption peak is caused by ozone in the stratosphere. This is why its influence on the spectral transmittance of the atmosphere is shown in Fig. 3.5 (which illustrates a sea-level case) by a broken line.

Both these windows in the infrared have been used for remote sensing studies of the surface of the Earth — and its cloud cover — from satellite altitudes. For example,

Table 3.1
Representative MRIR channels on American and Russian weather satellites. (Source: Barrett, 1970)

Channel	Waveband (μm)	Indication
TIROS		
1	6.0–6.5	Absorption by water vapour
1 (TIROS VII)	14.8–15.5	Absorption by carbon dioxide
2	8.0–12.0	'Atmospheric window' emission
3	0.2–6.2	Reflected solar radiation
4	7.0–30.0	Long-wave radiation to space
5	0.55–0.75	Reflected visible solar radiation
NIMBUS		
1	6.4–6.9	Absorption by water vapour
2	10.0–11.0	'Atmospheric window' emission
3	14.0–16.0	Absorption by carbon dioxide
4	5.0–30.0	Emitted long wave radiation
5	0.2–4.0	Reflected solar radiation
COSMOS		
1	0.3–3.0	Reflected solar radiation
2	3.0–30.0	Emitted long wave radiation
3	8.0–12.0	'Atmospheric window' emission

weather satellites like the American Nimbus and some members of the Russian Cosmos family have been equipped with infrared sensors designed to record radiation across, or, within, these two atmospheric windows. Let us enquire more deeply into the relationships between choice of window waveband and object of enquiry by noting in greater detail the appropriate sensors flown on recent Nimbus satellites (see also Chapters 9 and 10).

(a) *The Medium Resolution Infrared Radiometer (MRIR)*. This multichannel radiometer system records radiation simultaneously in five different wavebands in the visible and infrared portions of the spectrum. (See Table 3.1). One operates in the region from 10.0–11.0 μm. This region is the most transparent within the 8.5 μm – 14 μm window. (It will be remembered that the ozone absorption band affects terrestrial radiation recorded at satellite altitudes). The thermal radiation measured by MRIR is available both by day and night, a characteristic not shared by natural visible light, which is reflected, not emitted, radiation. The best resolution of the MRIR data is about 40 km.

(b) *The High Resolution Infrared Radiometer (HRIR)*. This operates solely in the narrow, clearly-defined window waveband, from 3.4–4.2 μm. Since the effective transmittance in this region is higher than in the broader window, HRIR has been designed to give a much sharper view of its target, the best resolution being about 3 km.

(c) *The Temperature Humidity Infrared Radiometer (THIR)*. This is a two-channel radiometer, recording through the broader window from 10–12 μm, and in a water vapour absorption channel centred on 6.7 μm. (Plate 3.1). The resolutions at the satellite subpoint are about 8 and 22 km respectively. THIR data have been available by direct read-out to local A.P.T. (Automatic Picture Transmission) stations around the world. A complication we have not mentioned before is that in the two principal infrared windows, clouds often obscure the surface of the Earth. Water vapour is transparent to radiation but aggregations of water droplets reflect and scatter incident radiant energy. These attenuating effects are so strong that even quite shallow clouds reduce the effective transmittance to zero. Therefore, where clouds are present in an atmospheric column their upper surfaces act as the effective radiating surfaces so far as sensors designed to exploit the atmospheric window wavebands are concerned. Whereas some HRIR data in particular have been used in the mapping of Earth surface temperature patterns, THIR data are intended to reveal patterns of water vapour distributions through the 6.7 μm measurements, and distributions of clouds through the 10–12 μm recordings. In cloudy areas the equivalent blackbody temperatures derived from the recorded radiation levels can be used to map the heights of the cloud tops: temperatures usually decline upwards through the troposphere. By statistical techniques cloud top temperatures can be translated into heights above the ground.

(d) *The Surface Composition Mapping Radiometer (SCMR)*. This measures radiation in the 8.4–9.5 μm and 10.2–11.4 μm (Plate 3.2) intervals of the broader window with a

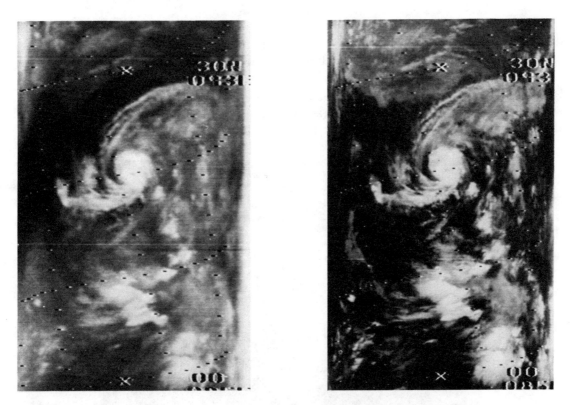

Plate 3.1
Significant differences are evident in these simultaneous views of the Bay of Bengal obtained from the THIR (11.5 μm and 6.7 μm channels) on Nimbus 5. The 11.5 μm image indicates surface and cloud top temperatures; the 6.7 μm image reveals the moisture content of the upper troposphere and stratosphere. (Courtesy, NASA)

600 m ground resolution. (Note that, once again, the ozone absorption band around 9.6 μm is avoided). These two channels yield different equivalent temperatures for the same target simultaneously, depending on its emissivity. Comparisons of the measurements made by the two channels of SCMR provide general indications of the types and variations of the mineral surfaces viewed from a satellite platform.

Before we leave for the time being the question of atmospheric window wavebands it should be understood that other important windows exist in addition to those we have discussed above. The best known and most widely exploited of them all is the waveband of visible light. Conventional photography, both colour and black and white, is hindered more by the common attenuating agencies of atmospheric water in droplet form and particulate matter than absorption by gases in the atmosphere. We shall discuss conventional photography in more detail later.

Other wavebands well worth exploiting for certain purposes are found in the microwave region of the electromagnetic spectrum. The opaque regions there are due to water vapour and atmospheric oxygen. The intervening windows have clear advantages over those in the infrared: they can be used even when weather conditions are quite severe. The relatively long wavelength rays in the microwave region (ranging from about 1mm to 1m) are able to penetrate thick clouds and even rain storms. Microwave sensing, whether active or passive, therefore has to some degree the 'all-weather' capability which other systems lack. At these longer wavelengths

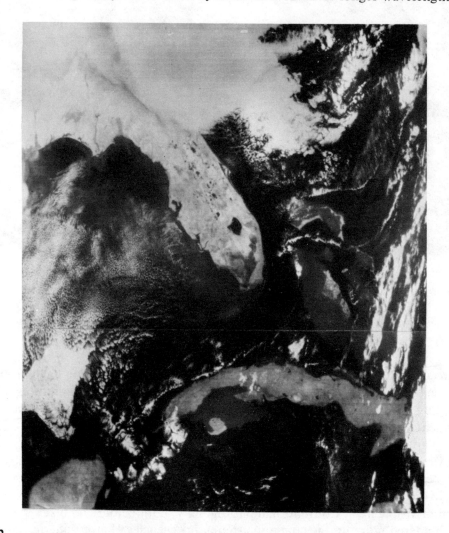

Plate 3.2
A Nimbus 5 Surface Composition Microwave Radiometer (SCMR) image (10.7 μm channel) showing Florida and Cuba on December 24th, 1972. Many fine-scale thermal features are evident, both over land and in the seas. (Courtesy, NASA)

scattering is relatively insignificant, and for normal atmospheric conditions absorption can be read for attenuation with little resultant error.

3.2.2 Recording reflected and/or emitted radiation

Not forgetting the significance of the altitude of remote sensing platforms, nor the special opportunities for remote sensing within atmospheric window wavebands, we can say that phenomena on the surface of the Earth can be investigated by several passive remote sensing means:

(a) Broad waveband sensing. Non-specific sensors can be used to integrate the energy from many wavelengths into a composite image. An example is the common ('panchromatic') camera-film combination. This records radiation across the visible portion of the spectrum within the upper and lower wavelength limits of the film emulsion.

(b) Narrow waveband sensing. Here radiation from the target is recorded only in a single selected waveband of the electromagnetic spectrum. We have seen how most objects reflect or emit energy over a broad range of individual wavelengths. There is, however, normally a peak wavelength at which the maximum amount of energy is being reflected or emitted. Under such conditions objects can best be differentiated from their backgrounds by measurements made at their radiation peaks.

(c) Bispectral sensing. Sometimes the location and identification of environmental phenomena is made easier by simultaneously recording radiation from the target in two non-adjacent wavebands. The data are then used comparatively. The Nimbus THIR and SCMR sensors perform such a type of operation.

(d) Multispectral sensing. A series of sensors arranged to operate at a number of very narrow bandwidths (often equally spaced across a selected waveband of the radiation spectrum) permit images or scan lines of spectral reflectance or emission from the target areas to be compiled. (Plate 3.3.). Much current work in remote sensing is now concerned with the compilation of spectral signatures, and their interpretation by comparison with 'fingerprint banks' of unique signatures known to be associated with particular objects or environmental associations. Unfortunately too little is known of the spectral responses of many targets in the natural environment. Many factors affect them, including temporal variations such as time of day and season of the year, Some multispectral scanners have employed as many as 24 separate channels, but the analysis of the signatures based on so many target responses is almost impossible without the aid of some mechanical or electrical back-up system. For manual analyses, sensors with as few as four carefully selected channels may suffice to provide maximum contrast for ready comparison.

Since we have already considered introductory examples of remote sensing by the first three of these four means we may fruitfully turn our attention finally to some examples of the fourth. As the analysis of multispectral signatures from the environment is hedged about by many difficulties and uncertainties we may with benefit consider first the nature of the responses from some individual constituents of that environment.

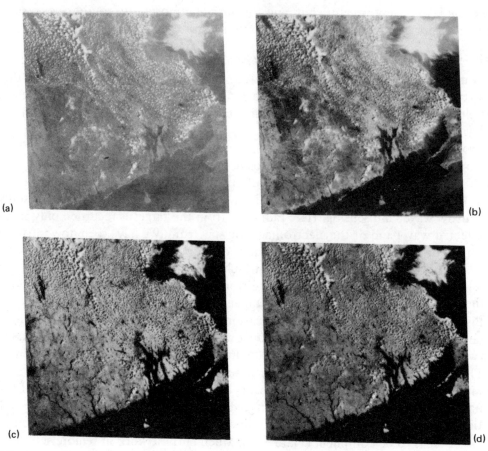

Plate 3.3
A set of ERTS-1 multispectral scanner film transparencies for New England July 28th, 1972, (a) Band 4 (0.5–0.6 μm); (b) Band 5 (0.6–0.7 μm); (c) Band 6 (0.7–0.8 μm); (d) Band 7 (0.8–0.9 μm).
(Courtesy, NASA)

3.3 Spectral signatures of the Earth's surface

3.3.1 Signatures of selected features

Rocks, the most fundamental constituents of the solid surface of the planet Earth, can be distinguished from each other under ideal conditions by their spectral signatures in the thermal emission region of the spectrum. We noted earlier that the emission spectrum from the Sun deviates in some details from the spectrum for a blackbody with the same mean surface temperature. Similarly many elements of the surface of the Earth have emissivities which vary both in temperature and frequency, and act more like grey bodies than black bodies. In Fig. 2.3 (b), illustrating the distribution of energy emitted by quartz

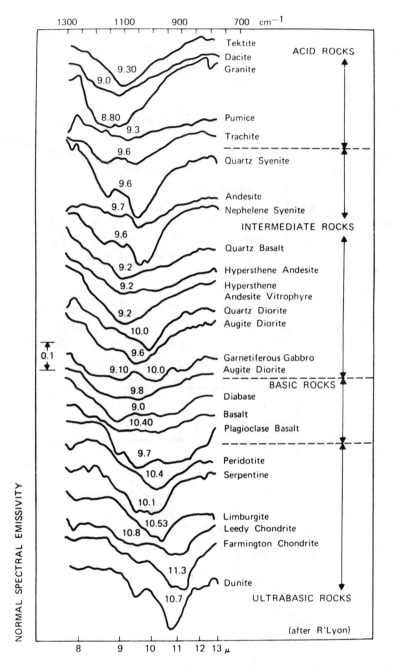

Fig. 3.6
Spectral signatures of rocks. Many types of rocks are differentiated from one another by the spectra of their radiation emissions in the thermal infrared. (Source: Laing, 1971; in *A Guide to Earth Satellites*, Fishlock, D., ed., Macdonald)

(SiO_2), there are wide differences between the two curves, especially around 9 μm and 20 μm. At these wavelengths incident energy is absorbed sharply by quartz, producing lower rates of emission here. Natural substances often behave more like perfect absorbers and radiators at certain frequencies than others revealing the so-called 'reststrahlen' or residual ray effect, when the actual emission curve is compared with the ideal. In the remote sensing of geological constituents of the environment basic rocks can be distinguished generally from acidic rocks by their signatures in the infrared, since the exact wavelength of the absorption peak varies from the one to the other (see Fig. 3.6). Fortunately such radiation differences can be observed even from satellite altitudes, since they fall within the broad atmospheric window waveband from about 8–13 μm. It has become possible to differentiate rough rock specimens in the laboratory by sole use of the spectrum of infrared energies emitted by the samples. The same method has been found equally applicable in the field and from the air. A simplified version of the multispectral method has been flown in aircraft using a spectral sensor with filters in the 8.0–9.5 μm and 10.0–12.0 μm wavebands. Further refinements may be expected in the future.

Similar differences between spectral signatures of different rocks have been described for the ultraviolet, visible and photographic infrared regions of the electromagnetic spectrum. One considerable cause of variability in the observed spectra from particular types of rocks is the water content of the samples. Another is the carbon dioxide content. Fig. 3.7 illustrates reflectance spectra of red sandstone surfaces under different moisture

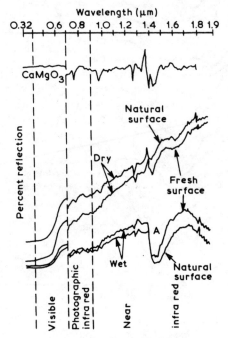

Fig. 3.7
Reflectance spectra of wet and dry, fresh and weathered (salt-encrusted) surfaces of a red sandstone.

conditions. It is notable that the percentage reflection from the wet surfaces is considerably less than that from the dry surfaces, especially around 1.4 and 1.9 μm which are strong water absorption bands. Clearly detailed 'ground truth' data must be obtained if these remote sensing data are to be correctly interpreted.

Broadening our discussion from rocks and weathered rocks to a wide variety of other environmental constituents, we may deduce from Table 3.2 that many types of soils, crops, and other active surfaces like ice and snow should be amenable also to multi-spectral identification from airborne or spaceborne sensor systems. These data, compiled during an Earth Resources Technology Satellite (ERTS) design study, represent aircraft and laboratory experience. Actual results from ERTS-I will be examined later.

Table 3.2
Aircraft and laboratory studies reveal differences in the percentage reflectances of different types of surface and crops. The four bands indicated here were those covered by the Multispectral Sensor (MSS) on ERTS-1.

Band	1 (0.5–0.6 μm)	2 (0.6–0.7 μm)	3 (0.7–0.8 μm)	4 (0.8–1.1 μm)
	Reflectance (%)			
Rock and soil materials and covers				
Sand	5.19	4.32	3.46	6.71
Loam 1% H$_2$O	6.70	6.79	6.10	14.01
Loam 20% H$_2$O	4.21	4.02	3.38	7.57
Ice	18.30	16.10	12.20	11.00
Snow	19.10	15.00	10.90	9.20
Cultivated land	3.27	2.39	1.58	(not given)
Clay	14.34	14.40	11.99	(not given)
Gneiss	7.02	6.54	5.37	10.70
Loose soil	7.40	6.91	5.68	(not given
Vegetation				
Wheat (low fertilizers)	3.44	2.27	3.56	8.95
Wheat (high fertilizers)	3.69	2.58	3.67	9.29
Water	3.75	2.24	1.20	1.89
Barley (healthy)	3.96	4.07	4.47	9.29
Barley (mildewed)	4.42	4.07	5.16	11.60
Oats	4.02	2.25	3.50	9.64
Oats	3.21	2.20	3.27	9.46
Soybean (high H$_2$O)	3.29	2.78	4.11	8.67
Soybean (low H$_2$O)	3.35	2.60	3.92	11.01

3.3.2 Signatures of complex environments

Some examples of multispectral imagery obtained from the ERTS-I spacecraft are shown in Plates 3.3 (a)-(d). One general feature of these pictures is immediately striking, namely the enormous complexity of their detail. It is quickly possible to make some comparative statements about such sets of images, for example the differing appearance of water

Table 3.3

Sources of variation in multispectral signatures of vegetation.
(Source: Polcyn *et al.*, 1969)

Illumination conditions
Illumination geometry (sun angle, cloud distribution)
Spectral distribution of radiation

Site environmental conditions
Meteorologic
Micrometeorologic
Hydrologic
Edaphic
Geomorphologic

Reflective and emissive properties
Spatial properties (geometrical form, density of plants, and pattern of distribution)
Spectral properties (e.g., reflectance or colour)
Thermal properties (emittance and temperature)

Plant conditions
Maturity
Variety
Physiological condition
    Turgidity
    Nutrient levels
    Disease
    Heat-exchange processes

Atmospheric conditions
Water vapour, aerosols, etc. (absorption, scattering, emission)

Viewing conditions
Observation geometry (scan angle, heading relative to Sun
Time of observation
Altitude

Multichannel sensor parameters
Electronic noise, drift, gain change
Accuracy and precision of measurements on calibration references and standards
Differences in spectral responses of systems

---

from one to another, but such statements are inevitably both qualitative and imprecise. More detailed and dependable conclusions necessitate a wide range of automatic and semi-automatic procedures for analysis and comparison. Later chapters are concerned with the more important of these methods. Unfortunately, of course, even these may not be completely successful, either at the level of recognising like objects or materials, or, much less, in identifying them. The first step towards a foolproof scheme must involve a consideration of the sources of variation in the signals from complex environmental assemblages of objects and materials. Table 3.3. lists the sources of variation which have been recognised in multispectral signatures of vegetation. The next step — evaluating the degrees of these influences on given sets of data — is fearsome to contemplate.

We may illustrate the complexity of the real world, and multispectral investigations of it, by reference to a programme carried out during March 1971 over Bear Lake on the border between Utah and Idaho. Plate 3.4 (a) shows the flight path of the instrument platform, on that occasion the NASA Convair 990 Airborne Observatory. In all, eight sensors were employed simultaneously to view the frozen lake and its surrounding snowfields and bare rock. The sensors are summarized in Table 3.4. One viewed in the infrared at 10 μm through the broad atmospheric window. The remainder were all microwave sensors, of which the 1.55 cm sensor was different from the rest in that it was a scanning device, whereas the others all viewed at fixed angles relative to the nadir angle of the aircraft.

Stripchart results from all the sensors are shown in Fig. 3.8. The plot for the 1.55 cm scanner is the average of the five central beam positions across the flight path of the aircraft. Considerable differences are evident from one curve to another. Some of these

Table 3.4

Characteristics of the radiometers employed in a programme of multispectral sensing over Bear Lake, Utah/Idaho, March, 1971. (Source: Schmugge *et al.*, 1973)

Frequency (GHz)	Wavelength (cm)	Pointing relative to nadir (deg.)	3dB beam width (deg.)	RMS temp. sens. (K)
1.42	21	0	15	5
2.69	11	0	27	0.5
4.99	6.0	0	5	15
10.69	2.8	0	7	1.5
19.35H	1.55	SCANNER	2.8	1.5
37V	0.81	45	5	3.5
37H	0.81	45	5	3.5
INFRARED	1.0×10^{-3}	14	< 1	< 1

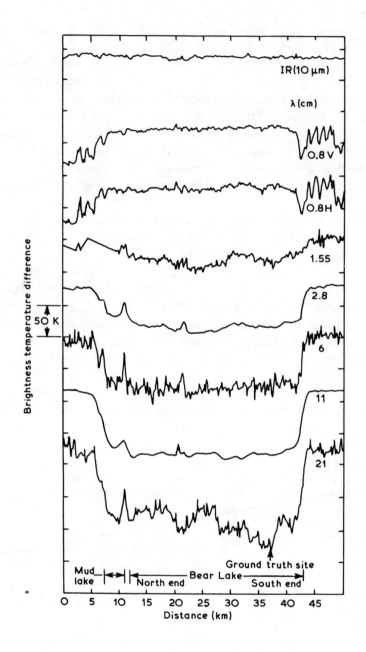

Fig. 3.8
Multispectral data obtained over Bear Lake, Utah/Idaho, March 3rd, 1971. H and V refer to the horizontal and vertical channels of the 0.8 cm radiometer, which viewed the surface at an angle of 45°. The remaining radiometers were nadir viewing. Each graph is related to the average temperature value at the frequency in question. (Source: Schmugge et al., 1973)

concern their general forms. For example, the curves of data from channels 2 and 3 are approximately the inverses of those from channels 5 and 6. Other differences relate to detail embroidered on the basic shapes.

One of the conclusions of this study was that at the longer wavelengths (especially 21 cm) the observed variability in brightness temperatures, especially over the lake, was related to thickness variations in the ice. Results indicate that the transparency of ice and snow is a function of wavelength, and that ice and snow depth may be assessed by these longer microwave emissions.

Plate 3.4 (b) (see colour plates) is the 1.55 cm microwave image of the pass over Bear Lake at an altitude of 3400m. The area it covers corresponds with part of Plate 3.3 (a). The outline of the frozen, variably snow-covered lake can be clearly seen. The low brightness temperatures of the frozen surface contrast sharply with the higher temperatures along the steep slopes of the eastern edge of the lake. Although a considerable amount of cloud was present over the target area when the flight was made, useful results were still obtained since microwaves, as we noted earlier, penetrate the clouds.

In conclusion it should be stressed that multispectral scanning, almost a science in its own right, is still at a very early stage in its development. Doubtless, data like those from the Bear Lake survey contain much more of interest and significance than we can recognize at present. The idea of investigating and evaluating the radiation characteristics of specific phenomena and environments simultaneously through a number of wavebands is highly attractive. For the most part, however, operational work today concerns itself with the simpler – but by no means simple – tasks of analysing integrated radiation, or radiation patterns obtained through one, two, three, or at most four, wavebands at one time. We know quite a lot about the basic theory of radiation reflection and emission. We need to learn much more about the highly complex reflection and emission patterns which are associated with real phenomena and situations.

References

Barnett, J.J. and Walshaw, C.D. (1974), 'Temperature measurement from a satellite', in *Environmental Remote Sensing; applications and achievements,* Barrett, E.C. and Curtis, L.F. (eds.), Edward Arnold, London, pp. 185-214.

Barrett, E.C. (1974), *Climatology from Satellites,* Methuen, London, pp. 19-74.

Fleagle, R.G. and Businger, J.A. (1963), *An Introduction to Atmospheric Physics,* Academic Press, New York.

Laing, W. (1971), 'Earth resources satellites', in *A Guide to Earth Satellites,* Fishlock, D. (ed.), Macdonald, London and Elsevier, New York, pp. 69-91.

Lockwood, J.G. (1974), *World Climatology: an environmental approach,* Edward Arnold, London, pp. 3-39.

Polcyn, F.C., Spansail, N.A. and Malida, W.A. (1969), 'How multispectral sensing can help the ecologist', in *Remote Sensing in Ecology,* Johnson, P.L. (ed.), University of Georgia Press, Athens, Georgia, pp. 194-218.

Sabatini, R.R. (ed.), (1972), *The Nimbus 5 User's Guide,* Allied Research Associates, Inc., Baltimore.

Schmugge, T. *et al.* (1973), *Microwave Signatures of Snow and Fresh Water Ice,* Publication No. X-652-73-335, NASA, Greenbelt, Maryland.

Sellers, W.D. (1965), *Physical Climatology*, University of Chicago Press, Chicago.

4 Sensors for environmental monitoring

In the discussion of radiation theory and radiation characteristics of natural phenomena some examples of sensors have already been introduced. In this chapter the sensors available for remote sensing studies will be examined in more detail in order to explain the advantages and disadvantages attached to each sensing system. It will be convenient to discuss the sensors in relation to the wavebands in which they can be applied. Two broad sensing categories can however, be identified at the outset — (a) photographic and (b) non-photographic (scanner dependent) systems (Fig. 4.1). Photographic systems operate in the visible and near infrared parts of the spectrum (0.36–0.9 μm) whereas non-photographic sensors can range from X-ray to radio wavelengths (see Fig. 2.2).

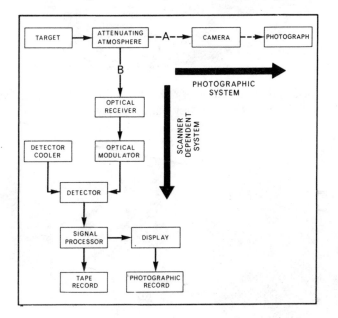

Fig. 4.1
Photographic and non-photographic (scanner dependent) systems. (Source: Curtis, 1973).

4.1 Sensors in the visible wavelengths

4.1.1 Photographic cameras
The photographic camera is a well known remote sensing system in which the focused

image is usually recorded by a photographic emulsion on flexible film base. There are three basic types of camera: framing, panoramic and strip cameras. The framing camera usually provides a square image with an angle of view up to 70°. The panoramic camera gives a very wide field of view which can extend to the horizons on either side,

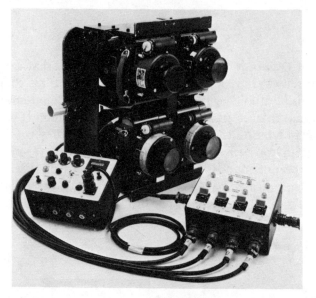

Plate 4.1a

Fairey multispectral camera system consisting of four Vinten cameras with lens focal length of 102 mm (4 in) loaded with 70 mm film. (Courtesy Fairey Surveys Ltd., Maidenhead, Berks.)

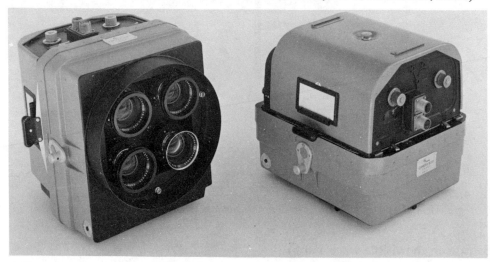

Plate 4.1b

International Imaging Systems (I² S) *multispectral aerial camera. Four Schneider Xenotar 150 mm or 100 mm lens. Film capacity 9½ in wide by 250 feet.* (Courtesy John Hadland (PI) Ltd., Bovington, Herts.)

with consequent distortion of the image. There are various types of panoramic camera systems and the commonest type uses a reciprocating, or continuously rotating, narrow field lens. The strip camera consists of a stationary lens and slit together with a moving film of width equal to, or greater than, the slit length. The film is moved across the slit at a speed which ensures no blurring of the image at the height flown (i.e. it is image-motion-compensated). The exposure is controlled by the slit width. Such cameras have been used mostly in low level sorties by high speed aircraft in military reconnaissance flights.

In the framing and panoramic cameras the exposure interval can be chosen so that a sequence of overlapping images can be obtained (Fig. 4.2 a,b). This overlap is essential if

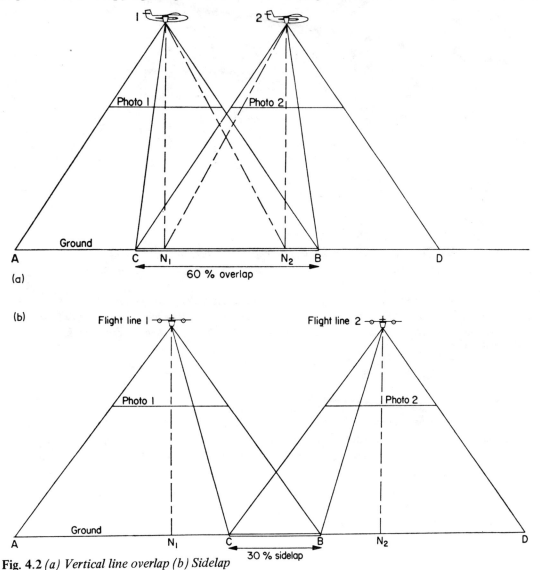

Fig. 4.2 *(a) Vertical line overlap (b) Sidelap*

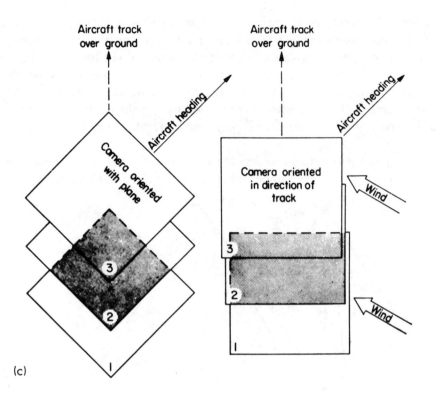

Fig. 4.2

Vertical line overlap photography.

(a) Overlap (endlap) photographic coverage is obtained by photography at time intervals which provide coverage AB at position 1 and CD at 2. Overlap BC is 60 percent and the nadir points N1 and N2 are within the overlap area.

(b) Sidelap (lateral) coverage is obtained by flight lines overlapping by 30 percent.

(c) The effect of crabbing. Overlap areas are shaded to show reduced overlap when camera is oriented with plane and the correction when camera is rotated to the direction of the aircraft track. (Source: Curtis, 1973.)

stereoscopic viewing is desired. In order to avoid gaps in the stereoscopic cover it is necessary to ensure that the camera is oriented in the direction of track of the aircraft or space platform. (Fig. 4.2c). The scale of the photograph obtained by a framing camera is dependent on the focal length of the lens and the height of the camera above ground. Mapping and reconnaissance cameras commonly use 6 inch lenses but 12, 3½, 3 and 1½ inch lenses are used in special camera systems. Where the 6 inch (150 mm) lens is used the calculation of scale of the photograph is a simple operation.

Example
Camera: Wild RC8. Universal Aviogon lens 6 in.
Height of camera above ground 10 000 ft.

$$\text{Scale of photograph} = \frac{\text{Focal length}}{\text{Ht. of camera above the ground}}$$

$$= \frac{6}{10\ 000 \times 12}\ \text{in} = \frac{1}{20\ 000}$$

The principal factors which limit resolution of the camera system are:
(a) Lens resolution.
(b) Film resolution.
(c) Film flatness and focal plane location.
(d) Accuracy of image-motion-compensation (I.M.C.).
(e) Control of roll, pitch, yaw, vibration.
(f) Optical quality of any filter or window placed in front of the lens.

Table 4.1

Sensitivity and resolution for Kodak aerial films. (Source: EMI, Department of Trade and Industry, 1973)

Type	Sensitivity (aerial film speed)*	Resolving power (lines mm^{-1}) at test object contrast:	
		1000:1	1.6:1
High definition Aerial 3414	8	630	250
Panatomic X 3400	64	160	63
Plus X Aerographic 2402	200	100	50
Double X Aerographic 2405	320	80	40
Tri X Aerographic 2403	640	80	20
Infrared Aerographic 2424	200	80	32
Aerochrome Infrared 2443	40	63	32
Ektachrome Aerographic 2448	32	80	40

* Aerial film speed for monochrome negative material is defined as $3/2E$ where E is the exposure (in metre-candela-seconds) at the point on the characteristic curve where the density is 0.3 above base plus fog density, under strictly defined conditions given in ANSI Standard PH2. 34-1969.
A A doubling of aerial film speed number denotes a doubling of sensitivity.

The definition of final image quality has conventionally been quoted in terms of resolving power. This is usually measured by imaging a standard target pattern (Fig. 4.3) and determining the spatial frequency (in lines per mm) at which the image is no longer distinguishable. Resolving power can also be determined by examining the contrast and distortion of the image of a sinusoidal grating object. This measurement, as a function of the spatial frequency of the grating gives what is termed the modulation transfer function (M.T.F.) which is commonly used to express the resolution characteristics of different films. Examples of the sensitivity and resolving power of Kodak aerial films are given in Table 4.1.

A summary of the characteristics of the different types of film available is given in Table 4.2 from which it will be apparent that the versatility of the camera sensing system is increased by the availability of different films with different spectral ranges, sensitivities and resolving powers.

Where images are required at different wavelengths multispectral photographs may be taken either by mounting several identical cameras with different film emulsions or using a special multispectral camera (Plate 4.1. a and b). In the latter several identical lenses record their separate images on a common film. The four lens system is the most common arrangement, but the Itek camera has nine lenses. An example of multispectral photography by the I² S camera system is given in Plate 4.2. This photograph was obtained with Infrared Aerograph 2424 film using four filters which provided images at 0.4–0.5 μm (1); 0.5–0.6 μm (2); 0.6–0.7 μm (3); and 0.7–0.9 μm (4). These separate images can be recombined and analysed as shown in Chapter 7.

The principal advantages of the photographic camera are its large information storage capacity, high ground resolution, relatively high sensitivity, and high reliability.

The main shortcomings of photographic surveys from spacecraft are that exposures can only be made in daylight, cloud obscures ground detail, and the photographic film cannot be re-used. It has been calculated that the film weight required to cover one year of operations of the Resources Satellite, ERTS-1 filming to obtain 60 metres ground resolution, is approximately 300 kg. On board processing would approximately double this weight, and improving the resolution to 20 metres would increase the film weight required to 3000 kg. This weight factor must be considered alongside the difficulty in transferring the information on the film to earth. One method of transfer involves processing the film on the satellite and electronically scanning it for picture transmission to ground stations by broadband telemetry. The other method is by ejecting the film from the satellite for recovery in the air or on the ground. Capsules of 39 kg (air retrieval) and 136 kg (ground retrieval) have been used in this way.

There is constant research aimed at improving the sensitivity and resolution of films. whilst reducing the thickness (weight) or films. Sensitivity has been approximately doubled each decade since 1850. In the last decade the film base has been reduced from 0.32 mm to 0.063 mm and in some satellite missions even thinner base materials have been used. As an example of the photographic camera facility now available one may

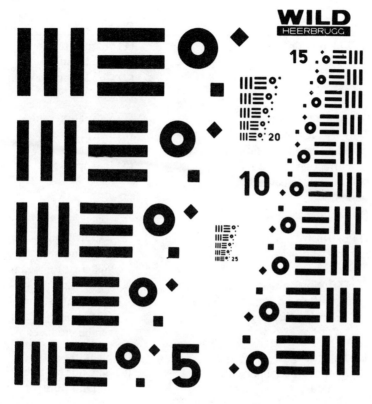

Fig. 4.3
Three line resolution test to determine resolving power. (Source: Smith, 1968)

note the characteristics of the S 190 Multispectral Photographic Facility as fitted in Skylab (Table 4.3).

4.1.2 The Vidicon camera

The development of the vidicon camera has provided satellites with a method for obtaining very high resolution pictures (0.35–1.1 μm) without the need for film replacement or converting a film image into a video signal. The principal of operation of a vidicon is shown in Fig. 4.4. The optical image is focused on the photo-conductive target and a charge pattern is built up on the target surface which is a replica of the

Table 4.2
Summary of film characteristics. (Source: Curtis, in Cruise and Newman, 1973, Academic Press).

Film	Disadvantages	Advantages
Panchromatic	(1) Limited tonal range	(1) Sharp definition (2) Good contrast (3) Good exposure latitude (4) Inexpensive
Infrared	(1) Limited tonal range (2) Very high contrast (3) Slight resolution fall-off (without correction for focal distance) (4) Difficult to determine correct exposure (5) Loss of shadow detail	(1) Vegetation evident (2) Tracing of water courses facilitated (3) Inexpensive
Colour	(1) Expensive (2) Special processing facilities necessary (3) Diffusion of image under high magnification and slightly less definition than panchromatic	(1) Excellent all-round interpretation properties due to good contrast and great tonal range (2) Good exposure latitude when used as a negative film (3) Good-quality black and white prints can be produced from negatives
False colour	(1) Expensive (2) Special processing facilities necessary (3) Diffusion of image under high magnification (4) Critical exposure (5) Low ASA rating (6) Less tonal range than colour (7) Duplicates expensive and difficult from positive film	(1) Sharp resolution (2) Superior rendering of vegetation and moisture

optical image. This positive pattern is retained until the electron beam again scans the surface and restores it to equilibrium by depositing electrons. An analogue television signal is thereby created which is then amplified. In the case of the Return Beam Vidicon (RBV) camera the signal is derived from the unused portion of the electron beam. This returns along the same path as the forward beam and is amplified in an electron multiplier. The RBV tube provides greater sensitivity at low light levels and higher resolution than the ordinary vidicon tube.

Although RBV cameras were installed in the Resources Satellite (ERTS-1) they

HSL UK 71 132   15 JULY 71   5,000 APPROX 100 MM   1 5 RUN   1 3 4 1

Plate 4.2

Multispectral photograph of part of Thetford, Norfolk. Apart from the differentiation of deciduous and coniferous trees Band 4 (infrared) also provides considerable additional detail concerning roof structures. Road details are better portrayed in Band 3 (red) at top left. A combination of Band 3 and 4 offer advantages in studies of urban landscapes. (Courtesy NERC)

failed to function satisfactorily and the two areas of limitation in their use at present are in focusing and in the stability of the optical system. The focus of the tube can fail due to lack of sufficient stability in the supply circuits over a period of a year in unmanned satellites. Another danger is that the optical system may become dislodged in the launch period.

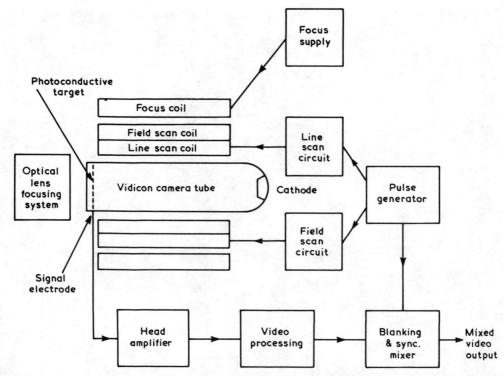

Fig. 4.4
Schematic diagram of a Vidicon Camera. The incoming radiation is focused on to the photoconductive target on which a replica of the optical images is formed. (Source: EMI, 1973)

Unlike the photographic film the slow scan vidicon photoconductive target may retain the previous image in vestigial form. Thus a 'priming' cycle is needed to allow complete discharge of the image and restoration of the target for the next exposure. In the ERTS-1 RBV camera this needed about 7 seconds and the complete cycle time could not be less than 11 seconds.

The vidicon camera can provide multispectral data either with a series of band pass filters on the optical system or with a number of vidicon sensors operating simultaneously to cover each band. This latter approach was used in ERTS-1 where three vidicon cameras covered different spectral ranges between 0.475–0.830 μm. The data transmission rate from this camera assembly is about 35 megabits s^{-1}.

Vidicon cameras have proved extremely valuable for producing real time and near real time imaging of the Earth's surface, especially in weather satellites. Experience with satellite borne television cameras carried in Tiros, Nimbus 1 and 2, the Ranger Moon shots, Apollo 8–16 and Mariners 1–9 has led to great improvements in vidicon systems. In the Mariner 6 and 7 systems good quality images were obtained of the surface of Mars. One camera (A) gave a ground resolution of 1 km and coverage of 900 x 700 km whereas a second camera (B) gave resolution of about 100 m in an area of 70 x 90 km.

Table 4.3
S190 Multispectral photographic facility (Skylab A). (Source: EMI, Department of Trade and Industry, 1973).

Six channels, six lenses (f/2.8, 150 mm), 70 mm film, 400 frames per cassette, 2.5 or 4 mil base, 57 mm square frame format, aperture stops f/2.8 to f/16 in ½ stop increments (± 1.5%). Shutter speeds 2.5, 5, 10 ms (± 2.5%), 4 ms synchronization. IMC: 10 to 30 mrad s^{-1} (± 5%). Boresighting − 60 arc s.

Sequence rate: 1 frame/2 s to 1 frame/20 s.

Dimensions, 570 x 610 x 460 mm; weight, 134 kg; power, 28 V d.c., 22 A peak.

Wavelength (μm)	Film type	Expected dynamic system resolution (line pairs mm^{-1})	Expected dynamic ground resolution (m)
0.5–0.6	Pan-X B & W	53	54
0.6–0.7	Pan-X B & W	63	54
0.7–0.8	1.R. B & W	26	107
0.8–0.9	1.R. B & W	26	107
0.5–0.88	1.R. false-colour	31	91
0.4–0.7	High resolution colour	53	55

System cost: approx. $400 000

4.2 Sensors outside the visible wavelengths

4.2.1 Infrared radiometers

The infrared radiometer measures the magnitude of the radiant flux incident on a detector. The radiation results from reflected and scattered solar radiation and self emission from the Earth's surface and atmosphere. Normally it is necessary to obtain measures of radiation at different wavelengths. This can be achieved by dispersing the incoming radiation by means of prisms, gratings, dichroic mirrors or filters. The dispersed radiation can then be measured in various wavebands using a number of detectors.

In order to obtain satisfactory coverage of the terrain and sufficient resolution it is usually necessary to adopt a system of scanning. The raster form of scanning (Fig. 4.5) is advantageous from the standpoint of data presentation but conical scan can also be employed. It will be apparent that scan rate must be related to the altitude and velocity of the sensor when assessing the performance of the system.

Detectors for use in scanning systems can be classed as thermal or photon detectors but Earth resource scanners invariably use photon detectors because of their greater sensitivity. A selection of photon detectors can be used to cover all the high atmospheric

Plate 4.3
Infrared Linescan Type 212 complete with cooling pack and control unit. (Courtesy Hawker Siddeley Dynamics Ltd., Hatfield, U.K.)

transmission wavebands from 0.4−14 μm. (Table 4.4.). Detector cooling is necessary to achieve satisfactory results for all detectors except for silicon (0.4−1.1 μm) and germanium (1.1−1.75 μm). Since long term stability of the detector cannot be assumed it is also necessary to provide some built-in calibration system. Calibrating signals are normally obtained using tungsten lamp, diffused solar radiation, or controlled blackbody temperature sources.

Since the maximum achievable radiance resolution can be limited by the degree of available detector cooling, considerable attention has been given to cooling systems. Cooling down to 100−77K is required for adequate resolution within the 8−13 μm waveband. Detectors used in aircraft usually use liquid coolant (liquid nitrogen) but for satellite multispectral scanners solid cryogen, closed cycle refrigerators or passive 'radiation to space' methods are used. With the solid cryogen a pre-cooled solidified gas is stored in a high performance dewar system and vented to space. Closed cycle refrigerators (e.g. Stirling) provide higher cooling capacity/weight/size ratios than solid cryogens but are demanding in terms of power requirements. Passive radiation cooling is an attractive proposition because of its lack of power requirements and long life. However, the cooling capacity of this system may be too low for some large detector arrays.

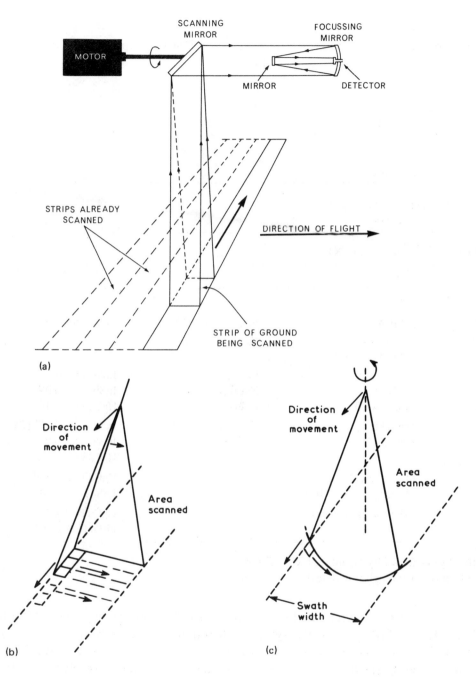

Fig. 4.5
Various forms of scanning (a) the linescan system (b) array scanning – side and forward motion (c) conical scanning.

Table 4.4

Optimum detector choice and waveband detectivities. (Source: E.M.T.I., Department of Trade and Industry, 1973).

Cooling

Thermoelectric cooling
300K _____ xK

Radiative cooling
300K _____ yK

Closed cycle engines and solid cryogens
300K _____ zK

Waveband (μm)	Temperature (K)			
	300	195	77	35

Detectivities (W^{-1} cm $Hz^{1/2}$)

Waveband (μm)	300	195	77	35
1.5 – 1.8	(1) InAs 4.5×10^{10} (2) HgCdTe (3) InSb 1.5×10^{8}	InAs 7.5×10^{10} HgCdTe	InAs 1.5×10^{11}	
2.0 – 2.5	InAs 6×10^{9} HgCdTe InSb 2×10^{8}	InAs 2×10^{11} HgCdTe 1.5×10^{10} InSb 3.5×10^{9}	InAs 3×10^{11} InSb 3.5×10^{10} HgCdTe 2×10	
3.5 – 4.1	HgCdTe InSb 3×10^{8}	HgCdTe 2.5×10^{10} InSb 6×10^{9}	InSb 6.5×10^{10} (PV) InSb 4.5×10^{10} (PC) HgCdTe 4×10	
8 – 10	(TGS 6×10^{8} (Thermistor 4×10^{8})		HgCdTe 3×10^{10} PbSnTe 7×10^{9}	GeHg 2.5×1
10 – 12	(Thermal detectors)		HgCdTe 2×10^{10} PbSnTe 9×10^{9}	GeHg 3×10^{1}

(Detectivities not degraded by $1/f$ noise. Field of view
$180°$ except for $8 - 12$ μm photon detectors where value is $60°$

Examples of radiometers used on satellite platforms have been given in Chapter 2 (MRIR, HRIR, THIR and SCMR systems). Infrared line scanning systems are also used extensively from aircraft and helicopter platforms. Infrared Linescan Type 212 (Plate 4.4) manufactured by Hawker Siddeley Dynamics has a temperature resolution of $0.25°$ C at $0°$ C background temperature, thus surface temperatures can be detected to an accuracy of less than $1°$ C. When the detector signals are processed to form an image the resultant thermal maps can provide great detail concerning temperature variations. Three examples

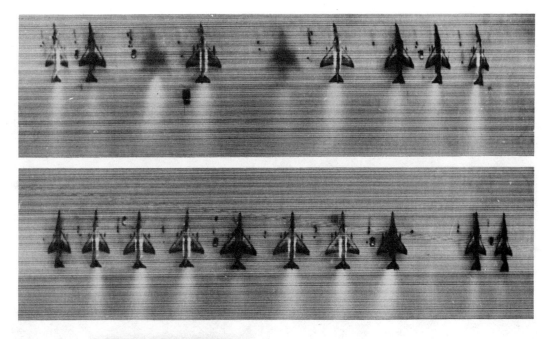

Plate 4.4a
Infrared linescan image of aircraft on a airfield. The heat from engines is shown in light tones. Note also the thermal shadows of recently departed aircraft at the top of the picture. (Courtesy Hawker Siddeley Dynamics Ltd., Hatfield, U.K.)

are shown in Plates 4.4 a,b where the heat from aircraft engines and the warmer areas in the shelter of hedgerow can be easily recognized. Infrared radiometers are also used by engineers of the Electricity Board to detect 'hot spots' in power lines of the electricity grid.

Thus infrared radiometry is much concerned with temperature mapping at different scales ranging from world wide patterns to small areas near an electricity pylon. It is also used as part of a multispectral scanning system in which the radiance at the infrared wavelengths are measured alongside the radiance in the visible wavebands. These visible and infrared wavelengths are affected by atmospheric conditions such as cloud and rain and consequently complete surface cover is often not possible. It has been estimated that, on average, any given area of the Earth is completely free from cloud for only 10–14 per cent of the time. In Europe the percentage number of days with less than 2/8 cloud varies from 20 in the northwest to 50 in southeast Europe. As a result repetitive observations of particular ground areas are difficult to achieve using visible and infrared scanners from satellites. It is for this reason that increasing interest is being shown in microwave radiometry.

Plate 4.4b
Infrared linescan imagery of land near Mark Yeo, Somerset. Note shelter effect of hedges showing in light tones of areas with higher temperature. Grazing animals are easily seen by reason of their high body temperature. (Courtesy Royal Radar Establishment, Malvern.)

4.2.2 Microwave radiometers

The measurement of radiation in the range 1–100 mm forms the basis of remote sensing by microwave radiometry. Although the microwave radiation intensities are much lower than those in the infrared, resulting in poorer temperature resolution, the longer wavelengths have the advantage that they allow sensing through cloud cover.

The microwave radiometer is made up of a directional aerial, a receiver (for selection and amplification) and a detector. A block diagram of a typical microwave system is shown in Fig. 4.12b and the applications of some radiometers are given in Table 4.5. The ground resolution achieved by microwave radiometers depends on the aerial size and the

Plate 4.5a
Radar imagery of the Malvern Hills. (Courtesy Royal Radar Establishment, Malvern)

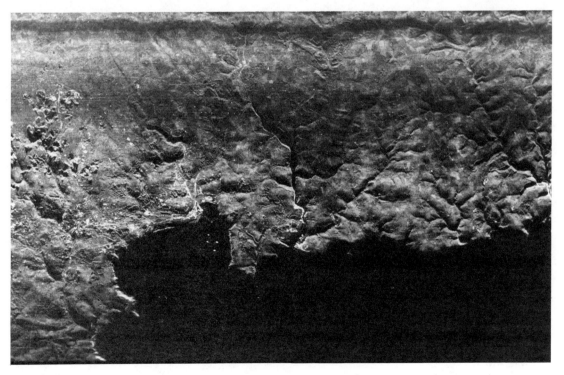

Plate 4.5b
Radar imagery of the coastal area near St. Austell. (Courtesy Royal Radar Establishment, Malvern)

orbiting altitude (Fig. 4.6). Since it is important that aerial size should be limited in order to avoid undue weight or distortion of the aerial array the lowest altitude orbit should be used for maximum resolution.

The principal characteristics of passive microwave radiometer (PMR) systems are, therefore:

(a) Low spatial resolution capability of the order of 1 km.

(b) Wide range in the observed surface emissivity.

(c) All weather capability.

The emissivity of an observed body changes with observation angle, polarization, frequency and surface roughness. An example of an object with low emissivity is a smooth water surface for which $\epsilon = 0.4$ when observed at normal incidence. Rough soil on the other hand has an emissivity close to unity. It is convenient when considering the measurements made by microwave radiometers to introduce the term brightness temperature, T_B. This is the temperature of a blackbody with the same emitted radiation as the physical object sensed by the radiometer.

Thus $T_B = \epsilon T$

where ϵ = emissivity

T = physical temperature (K)

Both natural and man made surfaces present a large range of brightness temperatures. Whereas for a blackbody the brightness temperature is equal to the physical temperature and ϵ is unity for all incidence angles, natural objects vary in brightness temperature and

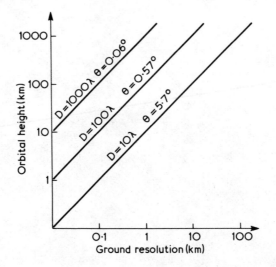

Fig. 4.6

The relationship between ground resolution, orbital height and effective aerial diameter, D. The aerial beamwidth is given by θ. (Source: EMI, 1973)

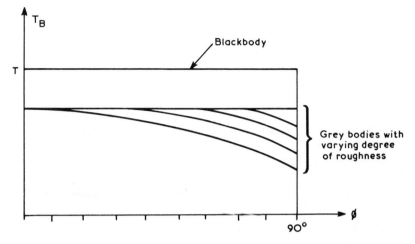

Fig. 4.7
Variations in brightness temperatures of black and grey bodies which are producing diffuse scattering.

in their response to incidence angles. Natural objects (grey bodies) show variations in emissivity depending on incidence angle, wavelength and the degree of surface roughness. Where surfaces are very rough the brightness temperature is independent of angle of incidence, but if roughness is limited T_B will become lower as the angle of incidence increases (Fig. 4.7).

An important question in relation to microwave radiometry concerns the depth to which ground properties contribute to the observed brightness temperature (i.e. the penetration depth). The penetration depth depends on the wavelength used and the dielectric properties of the material. Typical values of the penetration depth are 20 wavelengths for asphalt and sand and only 0.5–0.1 wavelengths for water. Thus wet materials provide little penetration whereas dry substances e.g. desert sands may show substantial (approximately 1 m) penetration. The wavelength dependence of the penetration depth makes profiling of the surface layer possible by using multifrequency radiometry.

Another important parameter characterizing the radiation properties of an object is the polarization i.e. the distribution of the electrical field in the plane normal to the propagation direction. As a rule the radiometer is only sensitive to the field in one direction. Blackbody radiation is completely unpolarized, but the emission of many natural features shows pronounced polarization effects which can be useful for identifying the nature of the feature.

The measured radiation of a surface feature depends on the following characteristics of the microwave system:
(a) Polarization direction of the radiometer.
(b) Observation angle related to the surface plane of the object.
(c) Frequency (wavelengths) of bands used.

Table 4.5

Microwave radiometry and its application to Earth resources surveys.
User requirements in hydrology and oceanography. (Source: Ohlsson, ESRO 1972)

Application	Spatial resolution (km)	Thermal resolution (K)	Frequency of coverage
1. Temperature monitoring			
Ocean wide, synoptic	150	1	bimonthly
Regional	20	1	weekly
Coastal and great lakes	5	1 (1)	on demand
Temperature gradient, ocean	150	(2)	
2. Soil Moisture			
Water content	0.2	5 (1)	daily to monthly
Gradient	0.2	(2)	daily to monthly
Horizontal moisture	0.2	5 (1)	on demand
3. Sea state	150	5 (1)	4 per day to daily
4. Oil film detection	0.1	10 (1)	on demand
5. Apportionment of water surfaces	1	10 (1)	daily to weekly
6. Ice/water boundaries	0.2	20 (1)	weekly
7. Snow			
Water equivalent of snow cover	0.2	(2)	weekly
Snow/land limits	0.2	5 (1)	weekly
Snow/ice limits	0.2	1 (1)	weekly
Snow temperature	0.1	1	on demand
Snow moisture	0.1	1 (1)	daily

(1) Required accuracy in measured brightness temperature, judged from
  data in the literature relating primary quantity to brightness temp.
(2) No satisfactory technique available; research needed.

The radiation received by the microwave system is also influenced by characteristics of the material which is being sensed. The important characteristics are:
(a) Electrical and thermal properties of the material.
(b) Surface roughness and size of the object.
(c) Temperature and its distribution in the body.
  At present the chief limitation of passive microwave sensing from space is the low

resolution of the system at orbit altitudes. However, the potential applications are considerable, as for example in the fields of hydrology and oceanography (Table 4.5). An example of an aircraft study is given on pp. 45–47 and some space applications have already proved to be of interest, such as the mapping of arctic sea ice.

4.2.3 Microwave radar

The microwave systems discussed so far have been *passive* systems, that is those receiving Earth emissions. It is possible, however, to devise an *active* sensing system in which waves are propagated near the sensor and are bounced on the Earth surface to be recorded on their return. This is the essence of the RADAR (Radio Detection and Ranging) system. It was first operated in the VHF radio band (30–300 MHz) but later improvements have led to the use of microwave radar (300 MHz–100 GHz).

Images of landscape derived from airborne side-looking radar resemble air photographs with low angle sun illumination in that shadow effects are produced (Plate 4.5 a,b). This enhances the impression of morphological relief in the imagery but also leads to some negative areas where radar shadows occur. The geometry of radar pictures is entirely different from that of conventional air photographs. This is because radar is essentially a distance ranging device which consequently produces lateral distortion of elevated objects as shown in Fig. 4.8.

The basic elements of a *Real Aperture* Side-Looking Radar System (SLAR) are shown in Fig. 4.9. A long aerial mounted on the airborne platform scans the terrain by means of a radar beam (pulse). The radar echoes are recorded by a receiver and processor so that a cathode ray tube (CRT) glows in response to the strength of each echo. The resultant CRT display is then recorded on film. These radars can be subdivided into six variants:

(a) Monofrequency. Monopolarized, transmitting a radar carrier polarized in one plane, usually horizontal.

(b) Multipolarized. This normally transmits a horizontally polarized electromagnetic wave but can receive horizontally and vertically polarized waves separately.

(c) Circular polarized. This transmits a circularly polarized wave and can receive right hand or left hand circularly polarized ground reflections.

(d) Multifrequency. This transmits two or more modulated radar carriers at widely differing wavelengths to show up differences in penetration and scattering from various surfaces.

(e) Panchromatic. This transmits a broad-band carrier in order to minimize diffraction effects (speckling).

(f) Polypanchromatic (a multifrequency panchromatic radar).

In most Real Aperture (SLAR) systems the resolution across the film record is up to ten times better than that achieved in the along track direction. The latter can only be improved by increasing the size of the antenna used. There is a limit, however, to the size of the antenna which can be mounted on aircraft and spacecraft. However the movement of the aircraft along its track can be used to form a long 'synthetic' aerial (Fig. 4.10). This system depends on a particular method of processing the Doppler information associated with the radar returns and is termed the Synthetic-Aperture System. The synthetic-aperture system is most easily understood by analogy with more familiar optical concepts. If one imagines that a plane wave of monochromatic light falls on a narrow slit the wave on the far side of the slit will diverge to form the familiar Fresnel zone pattern for a slit (Fig. 4.11a). The Fresnel zones on the screen are alternately dark and light in tone and are caused by wave interference patterns. The light wave amplitude that causes this observed variation can be shown to be both positive and negative. This represents the variation in the *phase* of the wave fronts at those points relative to the

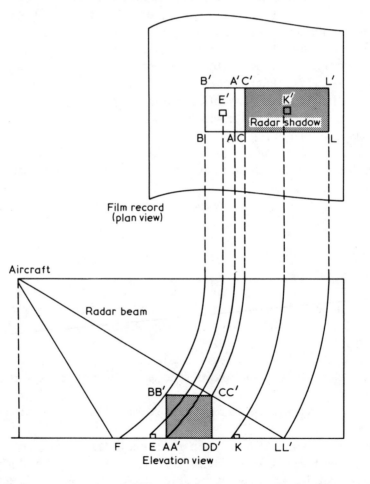

Fig. 4.8
Image distortions due to the directional and ranging characteristics of radar. (Source: Grant *et al.*, 1974)

phase of the wave arriving at the zero point i.e. the centre of the pattern on the screen.

It would be possible to place a film at the screen B and record the complete intensity image. Alternatively the same photograph could be obtained by traversing a narrow slit across the film (Fig. 4.11b) to expose it a piece at a time. If the phase of the waves could be added back it would then be possible to reconstruct an image of the slit A by shining a monochromatic (laser) light through the film record.

Now if one compares this example with the synthetic-aperture system radar the slit A can be regarded as a small part of the ground terrain. The slit C (Fig. 4.11b) is the recording system (CRT) for one item of information obtained by the radar receiver. The screen B (Fig. 4.11b) is the complete film record obtained from the CRT. display for the pulse striking the ground element represented by slit A.

If the image of the element (A) is to be reconstructed it is necessary to record both the phase and the amplitude of the signals received. This is achieved by carrying a very stable reference oscillator in the aircraft or spacecraft and comparing the phase of the return at each position of the sampling slit C with the phase of the stable oscillator. A

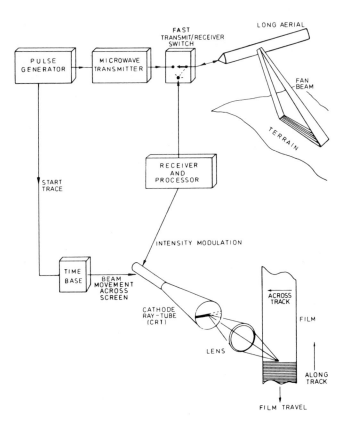

Fig. 4.9
Basic elements of a Real Aperture Radar System. (Side Looking Radar). (Source: Grant *et al.*, 1974)

radar with this type of stable oscillator is known as 'coherent' radar and it is an essential part of the synthetic radar system.

The advantages of side-looking radar compared with conventional aerial photography are:

(a) It has all weather and day/night capability.

(b) It is capable of measuring surface roughness and dielectric properties.

(c) It can penetrate into the surface layers of terrain (only small penetration in wet environments).

(d) It provides active illumination, with ability to select wavelength.

The disadvantages of the radar system are:

(a) Its poor resolution, although at satellite altitudes the potential resolution of a synthetic-aperture system is superior to that of an optical system.

(b) Its small scale.

(c) Its image distortion.

(d) Shadowing in areas of pronounced relief.

A particular type of radar system designed to obtain a measure of backscattered energy for various incident angles of the radar beam is termed the Scatterometer. The data obtained for each resolution element is presented in a graph of reflected energy versus incident angle. The variation of radar returns can be used by the scientist to determine roughness, texture, and orientation of the terrain. Different surface materials may also be identified.

Scattering signatures have been investigated experimentally for many materials and theory has been developed for statistical roughness parameters and electromagnetic parameters of various materials.

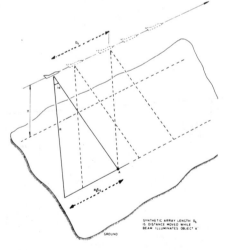

Fig. 4.10

Schematic diagram to show elements of a Synthetic Aperture Radar system. (Source: Grant *et al.*, 1974)

Operational scatterometers are used to gather data along a strip underneath the spacecraft or aircraft. Resolution capability is of the order of 1 m. On Skylab a narrow round beam was used which discretely shifted from the maximum angle to the nadir of the spacecraft as it moved over a given area.

4.2.4 Lidar (or laser radar)

The lidar is an active system similar to a microwave radar, but operating in that part of the spectrum comprising UV to near IR regions. It consists of a laser which emits radiation in pulse or continuous mode through a collinating system. A second optical system collects the radiation returned and focuses it on to a detector.

The lidar system is only effective in clear sky conditions due to the atmospheric absorption at the short wavelengths used. It is, however, a sensor of higher resolution than microwave radar. Three types of lidar are at present available: an altimeter type which

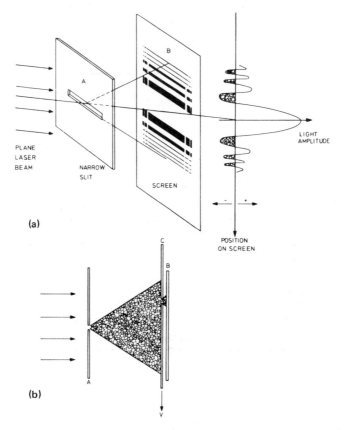

Fig. 4.11
(a) Fresnel zones appearing on screen and diagrammatic appearance of the phases of the waves. (b) Slit (c) moved across the face of the film (B). (Source: Grant et al., 1974)

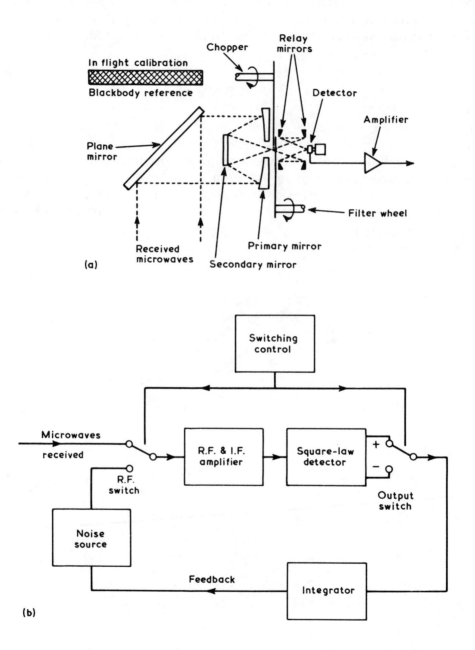

Fig. 4.12
Radiometer systems (a) Filter wheel radiometer (i.e. spectrometer) (b) Schematic diagram of microwave radiometer system.

can plot a terrain profile, a scanning type which can be used as a mapping instrument, and a third type employing spectroscopic techniques which may be used for mapping air pollutants.

4.3 Absorption spectrometry

In conclusion it may be noted that spectroscopic techniques can be applied to various types of sensors and where this is done the instrument is commonly referred to as a spectrometer. Spectrometers are essentially instruments used to determine the wavelength distribution of radiation. This can be done by dispersing the radiation spatially according to wavelength. The prism spectrometer relies on the dependence of the index of refraction of various prism materials on the wavelength of radiation entering the prism. The grating spectrometer achieves dispersion by diffraction and interference. One of the most popular grating spectrometer systems used today was devised by Ebert some sixty years ago. Scanning of the spectrum is achieved by applying a sawtooth type of oscillatory motion to the grating.

A rotating, circularly variable filter may replace the prism or grating and thereby gain an increase in the rate at which incoming radiation can be scanned. When a circular filter of this kind is used with a rotating beam modular (chopper) and an internal blackbody reference source, calibration of the spectrometer measurements can be achieved (Fig. 4.12a). Commercial circularly variable spectrometers are available with the capability of interchanging detectors. These instruments allow coverage of waveband frequencies in the range from 0.35–23 μm and can accommodate both cooled and uncooled detectors. Infrared spectrometers are often used to obtain atmospheric profiles — generally values of temperature and water vapour as a function of height. An infrared sensor which has demonstrated the feasibility of remote profiling of the atmosphere is the Satellite Infrared Spectrometer (SIRS) which uses an Ebert type grating spectrometer (See pp. 163). The more advanced Tiros Operational Vertical Sounder covers seventeen infrared bands from 3.7 μm to 29.4 μm. It is designed to measure temperature (1 K accuracy), relative humidity (10 per cent accuracy) and total ozone content (± 0.01 cm).

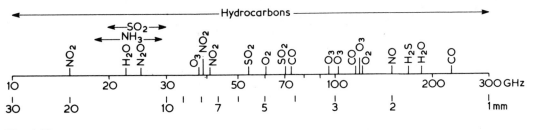

Fig. 4.13
Principal absorption frequencies of relevant gases in the microwave region. (Source: EMI, 1973)

Table 4.6

Applications of remote gas sensing. (Source: EMI, Department of Trade and Industry, 1973).

Discipline	Applications	Ground resolution desirable
Pollution	Determination of total vertical amount and vertical distribution of atmospheric gases (both man-made and natural). Can then determine (a) the global pollution background, (b) regional pollution dispersion and circulating patterns, (c) spatial and temporal variations of pollutants over urban, rural and ocean areas, (d) sink mechanisms, (e) atmospheric chemical mechanisms, (f) pollution flows across international boundaries.	L
	Mapping of small area pollutant sources, e.g. paper sand pulp mills, iron and steel plants, petroleum refineries, smelters and chemical plants, housing estates (where fuel is burnt for domestic purposes), busy road junctions (CO).	H
	Monitoring residues from large scale spraying of pesticides.	M
	Monitoring remote areas used for dumping of toxic materials.	M
Detection of Earth resources	Detection of metals from vapours emitted (e.g. arsenic, cadmium, zinc, sodium, mercury).	H
	Detection of micro gas seeps (especially NH_3) associated with natural gas fields.	H
	Detection of trace gases emitted from oxidizing ore deposits.	H
	Detection of iodine (iodine occurs in high concentrations in water in petroleum bearing strata, and is also associated with marine life which concentrates in primary fish food areas in oceans).	M
	Detection of chlorine (chlorine is indicative of bioproductivity in the oceans).	M
Other disciplines	Detection of volcanic activity by detection of SO_2.	H
	Determination of the composition of volcanic gases (enables predictions to be made of the composition of the molten mantle).	H
	Monitoring of volcanic emissions of SO_2 and H_2S to give warnings of eruptions.	H

Key. L — Low ground resolutions   (10 km)
       M — Medium ground resolution   (2 km)
       H — High ground resolution   (0.1—0.5 km).

The use of absorption spectrometry in remote gas sensing depends on the unique spectral signature of each gas. (Fig. 4.13). At present, spaceborne applications of remote gas sensing have been mainly limited to meteorological studies and to Martian atmospheric studies. However, the principles and instruments described could also be applied to the detection of pollutants and emitted gases and vapours from Earth sources. Up to the present no space instruments have been flown specifically for the detection of Earth pollutants or Earth resources. However, modifications of existing meteorological instruments and further technological development should enable some of the applications listed in Table 4.6 to be investigated.

References

Brock, G.C. (1970), *Image Evaluation for Aerial Photography*, Focal Press, London and New York.

EMI Electronics Ltd., (1973), *Handbook of Remote Sensing Techniques*, Department of Trade and Industry, London.

Grant, K., (1974), 'Side-looking radar systems and their potential application to earth-resources surveys', *ELDO/ESRO Scientific and Technical Review*, **6**, 117.

Savigear, R.A.G., Cox, N.R., Hardy, J.R., Hughes, J.F., Norman, J.W., and Roberts, E.H. (1974), 'Side-looking radar systems and their potential application to Earth-resources surveys', *ELDO/ESRO Scientific and Technical Review.* **6**, 137.

Smith, J.T. (ed.), (1968), *Manual of Color Aerial Photography*, 1st edn., American Society of Photogrammetry, p. 550.

Welch, R., (1972), 'Quality and Applications of Aerospace Imagery', *Photogrammetric Engineering,* **38**, 379.

5 Sensor platforms and sensor packages

Remote sensing techniques can be applied from different types of observation platforms and each platform, mobile or stable, has its own characteristics. Generally speaking, three platforms are of extreme interest for remote sensing: ground observation, airborne observation and spaceborne observation platforms (Fig. 5.1).

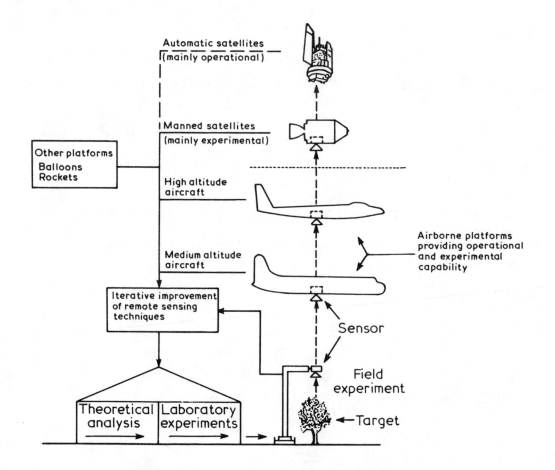

Fig. 5.1
Multi-platform remote sensing operations.

5.1 Ground observation platforms

A scientific field and laboratory programme is necessary to provide a good understanding of the basic physics of the object/sensor interaction. Measurements of physical characteristics such as spectral reflectance and emissivity of different natural phenomena are necessary in order to design and develop sensors. Normally fundamental studies are of two kinds:

(a) Laboratory studies where soils, plants, building materials, water bodies etc., are subjected to an external source of radiation and detectors of various kinds are used to measure reflection and emission spectra.

(b) Field investigations in which the spectral characteristics of surface phenomena of crops are investigated in different atmospheric conditions.

In field investigations the most popular platform to date has been the 'cherry-picker' platform (Fig. 5.2). These platforms can be extended to approximately 15 m. They have been used by the Laboratory for Agricultural Remote Sensing at Purdue University to carry spectral reflectance meters and photographic systems. These platforms are fre-

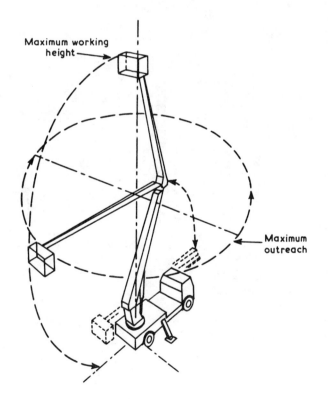

Maximum working height

Maximum outreach

Fig. 5.2
Mobile hydraulic platform of the 'Cherry-Picker' type. (Source: EMI, 1973)

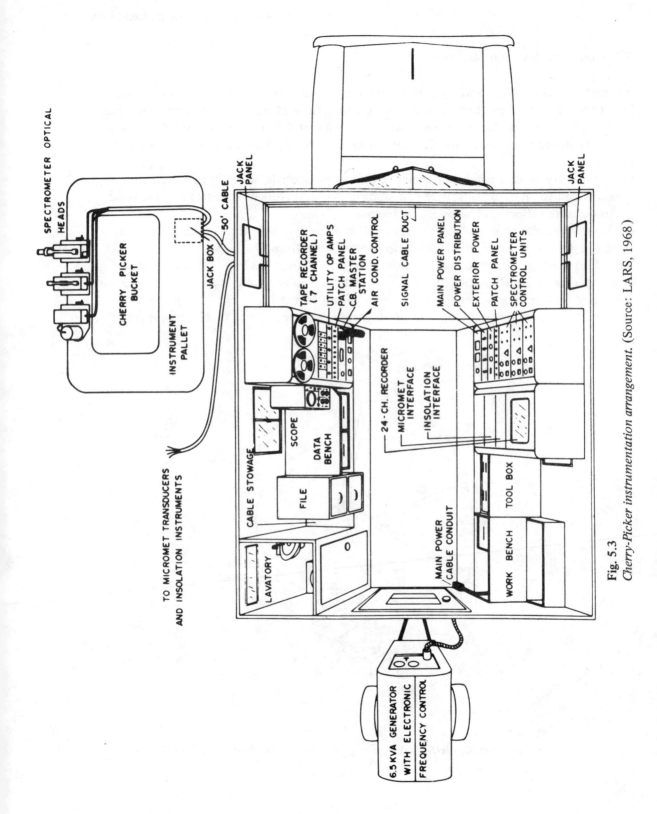

Fig. 5.3

Cherry-Picker instrumentation arrangement. (Source: LARS, 1968)

quently linked to automatic recording apparatus in field vans (Fig. 5.3). The limitation of this type of platform is that it is not a cross country vehicle and is limited to roads.

Portable masts are also available in various forms and can be used to support cameras and sensors for testing. The general problem with these masts is that of stabilizing the platform, particularly in windy conditions. They can also be mounted on vehicles as shown in Plate 5.1.

Towers are also available which can be dismantled and moved from one place to another. These offer greater rigidity than masts, but are less mobile and require more time to erect.

Plate 5.1a
An extending mast fitted to a Land Rover. The mast can be extended to 10 m height and is equipped with camera and photo-diode sensors for studies of crop reflectance. (Courtesy University of Bristol)

Plate 5.1b

Tower for supporting sensors. (Courtesy Hawker Siddeley Dynamics Ltd., Hatfield, Herts.)

Plate 5.2a

Balloon nacelle before flight. (Courtesy Société Européenne de Propulsion)

Plate 5.2b
Sensor equipment inside balloon nacelle. Photometer (black) and cameras in base plate. (Courtesy
Société Européenne de Propulsion)

5.2 Balloon platforms

Balloon platforms can be considered under two headings: tethered and free flying.
Tethered balloons were used for remote sensing observations (air photography) as long
ago as the American Civil War (1862) and very recently for nature conservancy studies
on beaches in Britain. Modern studies are, however, often carried out by free flying balloons
which can offer stable platforms up to very considerable altitudes. For example the
balloons developed by the Société Européenne de Propulsion can carry a photographic
nacelle (Plate 5.2) to altitudes of 30 km. This nacelle consists of a rigid circular base
plate supporting the whole equipment to which is nested a tight casing, externally
protected by an insulating and shockproof coating. It is roll stabilized and the inner

temperature is kept at 20° C and at ground atmospheric pressure. Standard equipment includes two cameras, multispectral photometer, power supply units and remote control apparatus. Its return to Earth is achieved by parachute, after remotely controlled tearing of the carrying balloon.

5.3 Aircraft platforms and remotely piloted vehicles

5.3.1 Aircraft platforms
Remote sensing from aircraft platforms has a fifty year background of using aircraft of different sizes and also helicopters or autogyros for photogrammtry. Many resource development and planning programmes employ aerial photography from altitudes of approximately 1500–3000m. These missions at low or medium altitudes are appropriate for surveys of local or limited regional interest. Precise flight times at short intervals and defined repetition rates over long time periods are only partly possible.

Aircraft platforms offer an economical method of testing sensors under development. Thus photographic cameras, line scanners, radar and microwave scanners have been tested over ground truth sites from aircraft platforms in many countries, especially in the NASA programme.

In order to obtain a greater areal coverage and to test atmospheric effects on sensors special high altitude aircraft have been used as plaftforms. These have often been developed out of military reconnaissance aircraft such as the American U2 plane (see Fig. 5.1). The altitudes attained by such aircraft are of the order of 15 km and a coverage of 100–400 km² per frame can be obtained.

5.3.2 Drones
A new type of platform is currently under development which can be used for remote sensing. This is usually termed the drone. It consists essentially of a centre body carrying the engine, propelling fan and fuel tanks. The tail section contains small wing structures and a tail plane together with control mechanisms which enable the engine and direction of flight to be controlled from the ground. The drone is, therefore, a remotely piloted vehicle capable of forward speeds of about 100 km h^{-1} and also hovering flight.

An example of a drone developed in Britain is the Short Skyspy (Plate 5.3). This vehicle is about 1.12 m in diameter and 1.37 m in overall height. It is a lightweight vehicle of 120 kg gross weight because the body is made from glass fibre reinforced plastic. The only major metal items are the engine and fan parts. The drone can carry a payload of approximately 20 kg which is mounted on the side of the vehicle.

The Skyspy has an endurance time of about 1.5 h, an operating altitude of about 500 m and a climb rate of 4 m s^{-1}. Aerodynamic controls are operated by separate servomotor systems which receive signals related to the attitude and position of the vehicle from sensors within the drone and from the ground. The drone sensors can provide information to maintain the drone in the attitude demanded by the ground control or by a self contained navigation system. It is envisaged that the drone will take off from a support framework on the ground or on a carrier vehicle. The movement and attitude of the drone in flight is directed by levers on a control console.

This platform was originally conceived for military reconnaissance but it is evident that it has a potentially useful role to fill in the testing and operation of remote sensing devices. These roles may include photography, infrared detection, radar observations and TV surveillance. Much will depend, however, on the ability shown to develop small payloads of comparatively low weight. The existing payload would not, for example, allow the drone to carry large multispectral cameras of the I^{2}S type.

The great advantage of such a device is that it could be accurately located above the area for which data was required. It is also an all-weather type of platform capable of both night and day observations.

5.4 High altitude sounding rockets

The dependence of remote sensing technique on the distance from the target can be

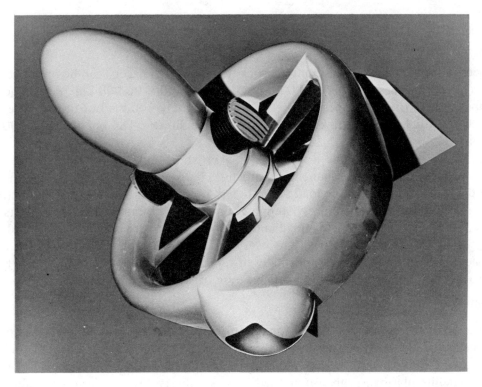

Plate 5.3
Model of Skyspy in flight. The payload is contained in the housing on the lower side of the vehicle.

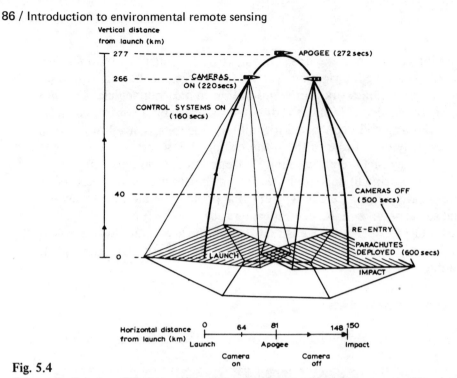

Fig. 5.4

Schematic diagram of Skylark rocket trajectory and photographic coverage. (Source: Savigear, *et al.,* 1974)

assessed by using high altitude sounding rockets. Synoptic imagery can be obtained from rockets for areas of some 500 000 square km (40 000–90 000 km² per frame). The apogee of existing European rockets is approximately 300 km but this may be extended to 400 km with 200 kg payload (c.f. Space Shuttle, 185 km orbit, 18 200 kg payload).

The Skylark Earth Resource Rocket (Fig. 5.4) is an example of such a platform. The rocket is fired from a mobile launcher to altitudes between 90–400 km. During the flight the rocket motor and payload separate and its sensors are held in a stable attitude by an automatic control system. By stepping the payload round 6 times the sensors can scan a 360° field of view. The payload and the spent motor are returned to ground gently by parachute enabling speedy recovery of the photographic records. The sensor payload consists of two cameras, normally of the Hasselblad 500 EL/70M 70 mm type. This rocket system has been used in surveys over Australia and Argentina.

5.5 Satellite platforms

Satellite observations are best suited for synoptic overviews of regional relationships at a small scale (35 000 km² per frame for ERTS-1). They also allow repeated complete coverage of the Earth and the required rate of repetition can, to some extent, be selected. Satellite platforms can also be used to make observations under comparable or identical conditions of illumination or for comparatively long periods.

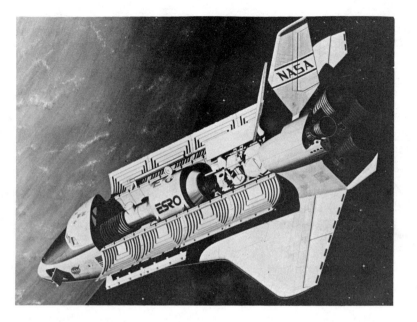

Plate 5.4
Artist's impression of Space Shuttle and Spacelab. Spacelab is being made in Europe and carries ESRO markings. (Source ESRO).

Two main types of satellite platforms can be distinguished namely automatic satellites, and manned satellites. In this book many satellite platforms are discussed in relation to particular studies, e.g. Gemini and Tiros. Two selected examples of satellite platforms will be discussed here – the Earth Resources Technology Satellite (ERTS) which is an automatic satellite, and Spacelab, which is a manned satellite platform to be launched in 1980, using the Space Shuttle (Plate 5.4).

The overall ERTS system is designed to make automatic observations using a payload consisting of return beam vidicon (RBV) camera system and Multispectral Scanner (MSS) (Fig. 5.5). The RBV system operates by shuttering three independent cameras simultaneously, each sensing a different spectral band in the range 0.48–0.83 μm. The MSS system is a line scanning device using an oscillating mirror to scan at right angles to the spacecraft flight direction. Optical energy is sensed simultaneously by an array of detectors in four spectral bands from 0.5–1.1 μm. There is a 10 per cent overlap of consecutive frames and an area of coverage approximately equal to that of the RBV images. When operated over a ground receiving station the data from the sensors are transmitted in real time (contemporaneously) to the ground receiving site and recorded there on magnetic tape. When the RBV and MSS sensors are operated at locations remote from a ground receiv-

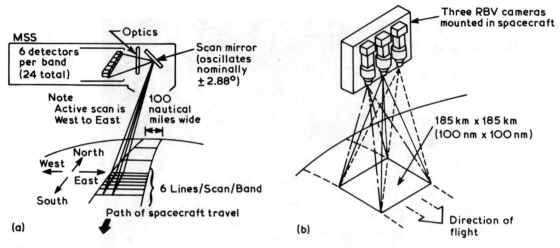

Fig. 5.5
(a) ERTS-1 multispectral scanning system — scan pattern. (b) ERTS-1 return beam vidicon scanning pattern. (Source: NASA)

ing station, two wideband video tape recorders (WBVTR) on board the space platform are used to record the video data. Each WBVTR records and reproduces either RBV or MSS data upon command.

The ERTS automatic satellite platform operates in a circular, sun synchronous, near-polar orbit at an altitude of 915 km. It circles the Earth every 103 minutes, completing 14 orbits per day and views the entire Earth every 18 days. The orbit has been selected so that the satellite ground trace repeats its Earth coverage at the same local time every 18 day period to within 37 km of its first orbit. A typical one-day ground coverage trace is shown in Fig. 5.6 (a) for the daylight portion of each orbital revolution.

Manned space platforms (e.g. Skylab and Spacelab) have certain advantages over automatic satellites. First, the greater payload capacity allows a complete instrumentation for all usable wavelengths within the electromagnetic spectrum. Equipment of heavy weight, large dimensions and high energy demands can be accommodated. Second, on-board experiments with different sensors or sensor settings can be made and in-flight decisions can be implemented. The main disadvantages of the Spacelab are relatively short (7–30 days) operation time, low repetition rates over a particular target area and inclinations of the orbit.

The Space Laboratory (Spacelab) will be carried into orbit in the payload bay of the Shuttle orbiter and remain attached to the orbiter throughout the mission (Plate 5.4). At the end of each mission the orbiter will make a runway landing back at the launch site, similar to that of a conventional aircraft (Fig. 5.7) The laboratory will then be prepared for its next mission. An important feature of the system will be its quick turn round capability of approximately two weeks.

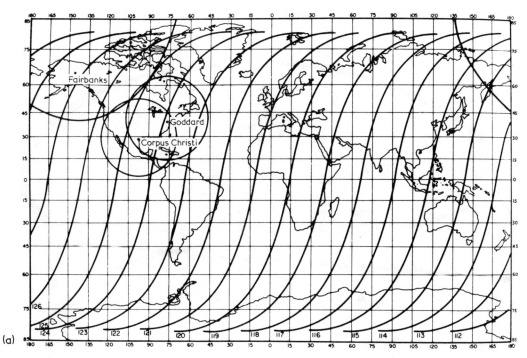

Fig. 5.6

(a) Sample subsatellite trace patterns for ERTS-1. (Source: Steiner, D., 1971, *Photogrammetria,* **27,** 211-251)

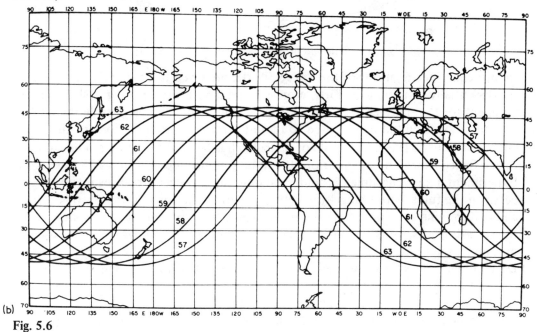

Fig. 5.6

(b) Sample subsatellite trace patterns for Skylab. (Source: Steiner, D., 1971, *Photogrammetria,* **27,** 211-251)

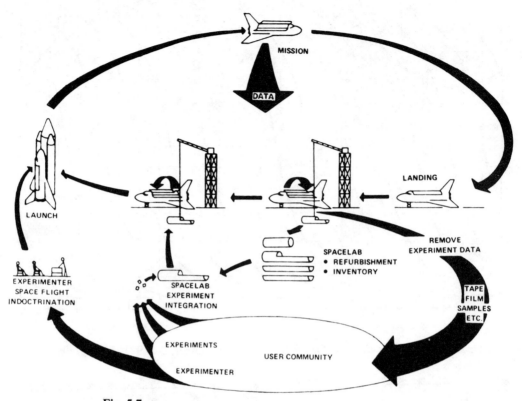

Fig. 5.7

Typical Shuttle-Spacelab profile. (Source: NASA/ESRO)

The principal specifications of the Space Shuttle are given in Table 5.1. The platform consists essentially of two parts. First, a pressurized laboratory offering a 'shirt sleeve' environment for the space crew and scientists. Second, a pallet area outside the laboratory which is unpressurized but can be used to mount certain items of equipment that can be remotely controlled. The general role of Spacelab has been outlined by studies in the European Space Research Organisation. It can be summarized as follows:

(a) Experimental missions aimed at understanding the physical processes involved between the full spectral range of electromagnetic radiation and the Earth's surface, and determining and converting the influences of the intervening atmosphere.

(b) Space testing of sensors and other equipment (e.g. large antennae) for later use in automatic satellites.

(c) Development of experiment methodologies and techniques e.g. comparison in space of different sensor types, combination of sensors or spectral bands, for use in future Spacelab missions or in automatic satellites.

(d) Development of on-board data processing, handling, compression and storage techniques.

Table 5.1
Summary of main user orientated Orbiter and Spacelab characteristics (Source:
Shapland D.J., Aerospace, 1974)

Parameter or function	Value or description
Payload weight/orbit altitude	29 500 kg/185 km. (on a due East launch). 11 340 kg limit for return.
Payload bay size	4.6 m diameter x 18.3 m length.
Orbitor size	Approximately the size of a DC–9 aeroplane.
Mission duration	7 days with extension possible to 30 days.
Crew	1–4 payload specialists.
Operating life	10 years/at least 500 flights.
Pointing accuracy	$\pm 0.5°$ (selectively as low as $0.1°$).
Communications	Down link via orbiter and Tracking and Data Relay Satellite (TDRS) provides up to 50 mbps digital or 6 MH$_2$ analog and video. Telemetry 20 kbps interleaved with orbiter data. Up link via orbitor 2 kbps command.
Electrical power	4 kW average } unregulated dc 9 kW peak } 26V to 32V dc 115 V/200 V/400 H$_2$ 3-phase.
Environmental control	Normal atmosphere N$_2$/O$_2$. Air temperature 18–26°C. Relative humidity 30% to 70%. Cleanliness 5 μm filtering.
Data management	Separate sub-system and payload computers 16 bit (fixed) and 32 bit (fixed and floating point) operations. 500 K ops s^{-1}. 64 K 16 bit word memory. Alpha numeric display and keyboard.
Launch sites	Eastern Test Range (Florida). Western Test Range will be employed for polar orbits.
Turn around time (Shuttle)	14 calendar days.
Safety	At least fail safe principle, intact abort at any time.
Operational data	Initial operation capability 1980 (launch rate of about 10 per year).

(e) Investigations into the role of man in manned space platforms used for earth
  survey missions.

The payload of Spacelab is likely to consist of three main groups of sensors. These
are facility, standard support and experimental sensors. Facility sensors are considered

as basic sensors common to a wide range of mission types and users. Typical examples of facility sensors are multispectral scanners and the active and passive microwave sensors. Standard support sensors and equipment are any additional sensors required to support the main sensor payload. Examples of these are metric camera, multispectral cameras, tracking telescope, precision altimeter, precision non-scanning spectroradiometer, data receiving and display equipment and consoles allowing the crew to operate and monitor sensors during the mission. Experimental sensors could be extensions or modifications to the basic facility payload, or new sensors specially developed for space application. New sensors might include laser radar, imaging spectrometer, and polypanchromatic side-looking radar.

It is possible to identify optimum groupings of sensors in which each module would comprise one facility/support sensor plus the necessary supporting sensors. Three sensor complements using different core sensors are shown in Table 5.2.

5.6 The relationships between ERTS, Skylab and the Space Shuttle

NASA's Goddard Space Flight Center (GSFC) began a conceptual study of the feasibility of Earth resources satellites in 1967. A sequence of six satellites with designations A,B,C, D,E and F (after launch becoming 1,2,3,4,5 and 6) were planned. ERTS-A which became known as ERTS-1 after launch in 1972 was regarded as essentially a feasibility experiment

Table 5.2
Optimum sensor groupings about a single major sensor
(Source: ESRO, 1974)

Payload cores or modules	Core sensors	Support sensors
Multispectral scanner core	Multispectral scanner	Spectrometer Cameras 1R radiometer 1R scanner (1) Laser profiler
Synthetic aperture radar core	Side looking radar	Cameras Scatterometer
Passive microwave radiometer core (multifrequency)	Passive microwave radiometer	Cameras 1R radiometer 1R scanner Scatterometer

(1) If not included in the multispectral scanner.

Table 5.3

ERTS–A/B and Skylab sensors and their spectral characteristics. (After Steiner, D., *Photogrammetria*, **27**, 1971)

Spacecraft designation	Type of sensor	Spectral sensitivity range	Comments
ERTS–A/B RBV	Return beam vidicon TV camera		
	no. 1	475–575 nm	
	no. 2	580–680 nm	
	no. 3	690–830 nm	
MSS	Multispectral point scanner		
	channel 1	500–600 nm	
	channel 2	600–700 nm	
	channel 3	700–800 nm	
	channel 4	800–1100 nm	
Skylab S–190	Multispectral camera cluster		Type of film:
	no. 1	500–600 nm	B & W: Panatomic-X
	no. 2	600–700 nm	B & W: Panatomic-X
	no. 3	700–800 nm	B & W: IR Aerographic
	no. 4	800–900 nm	B & W: IR Aerographic
	no. 5	500–880 nm	Colour: IR Aerochrome
	no. 6	400–700 nm	Colour: SO–242*
S–191	Infrared spectrometer	0.4–2.4 μm and 6–16 μm	Spectral resolution 2×10^{-2} μm to 2.5×10^{-1} μm (lower at longer wavelength)
S–192	Multispectral scanner		
	channel 1	410–460 nm	
	channel 2	460–510 nm	
	channel 3	520–556 nm	
	channel 4	565–609 nm	
	channel 5	620–670 nm	
	channel 6	680–762 nm	
	channel 7	783–880 nm	
	channel 8	980–1080 nm	
	channel 9	1.09–1.19 μm	
	channel 10	1.20–1.30 μm	
	channel 11	1.55–1.75 μm	
	channel 12	2.10–2.35 μm	
	channel 13	10.2–12.5 μm	
S–193	Microwave system (passive + active): radiometer, scatterometer, altimeter	13.8–14.0 GHz	
S–194	L–band microwave radiometer	1.4–1.427 GHz	Centre frequency of 1.4135 GHz: bandwidth 27 MHz

* New film developed by Kodak, with high resolution at low contrast.

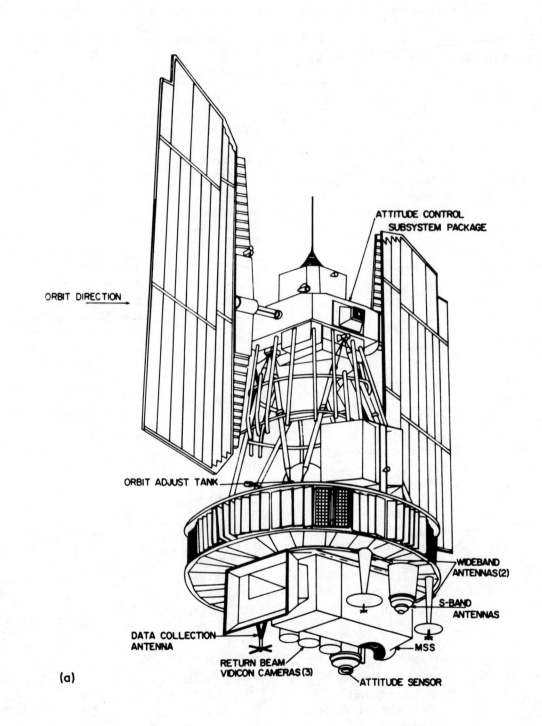

ORBIT DIRECTION

ATTITUDE CONTROL
SUBSYSTEM PACKAGE

ORBIT ADJUST TANK

WIDEBAND
ANTENNAS(2)

S-BAND
ANTENNAS

DATA COLLECTION
ANTENNA

MSS

RETURN BEAM
VIDICON CAMERAS(3)

ATTITUDE SENSOR

(a)

Fig. 5.8 (a) General configurations of ERTS-1

with instrument testing being a major objective. It was hoped that some useful terrain and resource data might be obtained at the same time and the results that have come from ERTS-1 have probably exceeded most expectations. ERTS-B was launched early in 1975 and it was planned to include a thermal infrared channel in the payload for this satellite. This channel was not available and so the sensors were similar to those included on ERTS-1. Subsequently to its launching the programme was renamed, ERTS-1 becoming Landsat 1 in retrospect, and ERTS-2 becoming Landsat 2.

In parallel to the Landsat programme NASA had the idea of developing an Orbital Research Laboratory and this concept was subsequently built into the Apollo programme. This project was named the Skylab project and the general configuration of Skylab is shown in Fig. 5.8 (b). The Skylab mission of 1973 carried a three-man crew and the observations were coordinated with the collection of ground data on test sites and aircraft underflights. Skylab can be regarded as a step in the development of the Space Shuttle and its successful completion was marked by the accquisition of detailed imagery of many parts of the world. It should be noted, however, that due to its lower orbit its coverage was limited to 50° N to 50° S only (see Fig. 5.6 (b)).

The ERTS-1 and Skylab sensors and their spectral characteristics are compared in Table 5.3 and the wider range of sensors available on Skylab is readily apparent.

In this discussion examples have been taken from the American NASA programme but it must be emphasized that a whole range of space platforms has also been

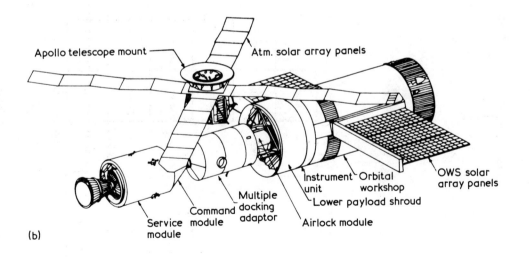

(b)

Fig. 5.8   (b) General configurations of Skylab platforms.

Table 5.4

Table 5.4 *Summary schedule of manned space flights.* (After Plevin, 1973)

1960	1961	1962	1963	1964	1965	1966	1967	1968	1969	1970	1971	1972	1973	1974
	Mercury Sigma 7								Apollo 12					
	Vostok 4								Soyuz 8					
	Vostok 3								Soyuz 7					
	Mercury Aurora 7								Soyuz 6					
	Mercury Friendship 7	Vostok 6							Apollo 11					
	Vostok 2	Vostok 5			Gemini 6			Apollo 8	Apollo 10		Soyuz 11 Salyut 1			
	Vostok 1	Mercury Faith 7			Gemini 7	Gemini 12		Soyuz 3	Apollo 9		Apollo 15			
					Gemini 5	Gemini 11		Soyuz 2	Soyuz 5	Soyuz 9	Soyuz 10 Salyut 1	Apollo 17	Soyuz 16	
					Gemini 4	Gemini 10	Soyuz 1	Apollo 7	Soyuz 4	Apollo 13	Apollo 14	Apollo 16	Skylab 1–4	
					Gemini 3	Gemini 9								
				Voshkod 1	Voshkod 2	Gemini 8								

developed in the U.S.S.R. (Table 5.4). The Soyuz Spacecraft (1–16) have shown progressive modification and the modular build up of an orbital space station (Salyut) is a major part of the Soviet space effort. In this respect the programmes of the two principal space nations show similarities which are discussed further in Chapter 17.

References

Cassinis, R., Bodechtel, J., Curtis, L.F., Long, G., de Loor, G.P., Strubing, K., (1973), *Views of the ESRO-PART Group on Earth Resources Missions in the Post-Apollo Programme,* ESRO/PA/R 102, Paris, p. 58.

Plevin, J., (1973), 'Earth resources and manned space systems – the role for spacelab', *The Implications for European Space Programmes of the Possibilities of Manned Missions; IV Earth Resources,* ESRO Summer School, 1973.

Smolders, P., (1973), *Soviets in Space,* Lutterworth Press, London.

Spurr, S.H., (1960), *Photogrammetry and Photo-Interpretation, Ronald Press, New York.*

Collecting *in situ* data for remote sensing data interpretation

6.1 Introduction: The need for *in situ* data

The use of remote sensing techniques demands that there should be some method of calibrating the sensors in use or checking the accuracy of the data obtained. For example where infrared thermal line-scan equipment is being used it is useful to have some temperatures measured on the surface to check the accuracy of the data. Also some equipment is designed to detect relative differences in temperature and display them to advantage i.e. they are self regulating systems which alter the scale over which measurement is made according to the dynamic range in the target area. In such cases it is essential to measure temperature at points within the target area in order to obtain some absolute values from the sensor data.

There is also a need for ground truth to check the accuracy of interpretations made on the basis of sensor data. For example if sensor data is being used to identify agricultural land use it is necessary to know the actual ground condition of a sample population of fields in order that the per cent accuracy of the identification can be determined.

It is common practice to use the term 'ground truth' for observations made on the surface of the Earth in relation to remote sensing studies. In many ways, however, the term is unsatisfactory because the measured variable is often in water, ice or air rather than on a ground surface. For this reason some workers prefer to use the term 'surface truth' instead. Even this term is not wholly satisfactory because it is sometimes difficult to define the measured 'surface' very precisely. As an example, it is often necessary to monitor meterorological factors such as windspeed, solar radiation, rainfall, atmospheric humidity and cloud cover in order to assess the effects of atmospheric conditions on sensor performance. Some of these measurements may be made close to the ground (microclimatic observations) others may be obtained from standard meteorological screens (mesoclimate observations) but some may be recorded at considerable heights in the atmosphere by radio-sonde techniques.

Similarly a wide variety of methods are employed for measuring *in situ* data in water bodies. Oceanographic observations may include measurements of sea temperature, salinity, wave motion (height and wavelength), and biological content as well as meteorological conditions. These measurements are made from a variety of platforms including weather ships, coastal protection vessels, automatic data collection buoys and coastguard stations.

In estuaries and rivers additional factors such as suspended sediment load, biological oxygen demand, water reaction and pollution are often recorded. These observations may be made from boats or by automatic samplers used from the banks.

In this chapter we shall draw examples from ground surface studies but the general problems of data collection such as sampling and selection of sites for observations are relevant to other types of surface such as water or ice. Also the data transmission methods are essentially the same whether one is observing water or land.

6.2 Selection of ground truth sites

The location of areas for ground truth collection may be decided on the basis of a number of criteria. These include study objectives, sample size satisfactory for statistical purposes, repeatability and continuity of the experimental study, access to the study area, availability of existing ground truth data for the area, personnel, equipment resources and the orbit characteristics of the space platform.

The shape of the ground data collection area is dependent on statistical requirements and speed of access. The allocation of sample plots in a data collection area is made more statistically efficient if the area can be stratified into relatively homogeneous areas. For stratification to be useful, strata boundaries should separate areas where within class variance is less than between class variance. The number of survey plots can then be estimated by the method of proportional allocation. Thus, the shapes of homogeneous areas may influence the shapes of ground survey areas. In addition, the sampling technique used in the ground observations (grid, area, line) may influence the shape of the area, for example line sampling along existing road networks may be preferred to block area sampling.

The size of the ground data collection area will be affected by study objectives, statistical considerations, scale of the ground phenomena, angle of view of the sensors and time factors. Where sensor testing is the objective it is desirable to select the smallest ground truth area which allows detailed ground monitoring by equipment and personnel over a wide range of ground/atmospheric conditions. Many small sites may be necessary if it is desired to check on the validity of a spectral signature for a particular surface condition. An example of such a site is that used by one of the authors at Long Ashton near Bristol (Plate 6.1). The total size of the site is 120 hectares within which smaller areas of approximately 2 hectares were often monitored.

Where relationships are sought between the sensor response and particular surface conditions (e.g. a particular crop such as grass) it is necessary to obtain sufficient samples of the crop (generally more than 30) to allow statistical tests to be carried out. The size of area which will provide such samples must be determined.

Time constraints affect the size of ground data collection areas through (a) quantity of data required and resources available for collection, and (b) the rate of change in ground environmental conditions. For example, where evaporation rates are high it may be necessary to monitor changes in soil moisture content frequently (several times a day)

Plate 6.1 *Ground truth site, Long Ashton. (Note the soil moisture measuring equipment in foreground).*

whereas in conditions of low evaporation loss the soil moisture change may be slow and fewer observations are required in a given time.

The prime objective of ground data collection is to provide a contemporaneous record of ground conditions at the time of imagery. In practice it is difficult to obtain synchronous data for more than a small area or selected sample sites. The aim, however, is to obtain sample ground truth data within a short time of the acquisition sensor data. In planning

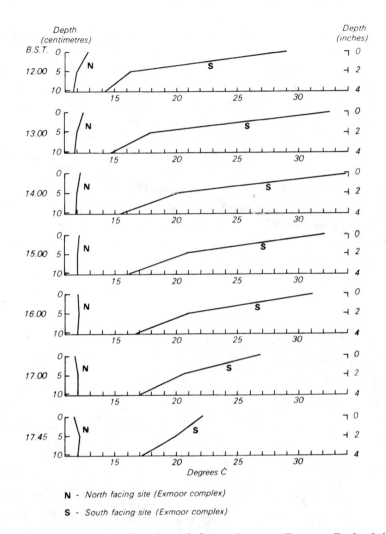

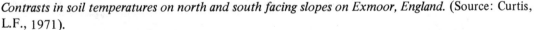

Fig. 6.1
Contrasts in soil temperatures on north and south facing slopes on Exmoor, England. (Source: Curtis, L.F., 1971).

ground data collection, special attention should be given to the rate of change of the variables to be observed. These variables can be categorised as transient or non-transient. Data recording of transient features (e.g. crop stage, leaf cover, windspeed, surface moisture) must be near synchronous. Recording of non-transient features (e.g. slope, aspect, soil texture) can be carried out prior to, or after, the sensing mission.

As an example of rate of change in a transient surface condition one may note that data for barley in Nottinghamshire showed that mean per cent leaf cover increased from 18 to 40 in a period of 8–10 days in the first half of May. An example of soil temperature variation on slopes of an Exmoor Valley (Fig. 6.1) shows that frequent observations are necessary on south facing slopes but the rate of change on north facing slopes is much less.

The range of ground data required varies according to the nature of the study. For example in crop studies it is necessary to record crop type, stage, height, colour, disease types, weed species, husbandry (ploughed, harrowed, drilled, rolled, wheelings), grazing method, livestock and per cent crop cover. In studies of soil conditions the ground truth data normally includes records of soil phase (or soil series), soil moisture, soil temperature, soil texture, structure, stoniness, organic matter content, soil colour, and bulk density.

The morphology of ground truth sites is also important since gradient, slope form and aspect may have significant effects on sensor data. Terrain classification and evaluation techniques are now well developed and they can be used for the recording of site morphology. Morphology can normally be recorded by field survey and analysis of contour maps before the sensing missions take place.

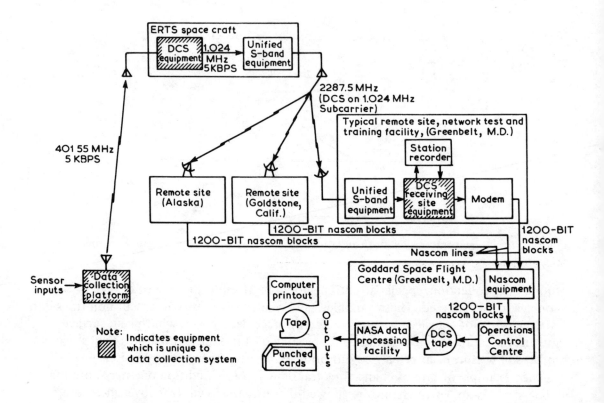

Fig. 6.2
Data collection system for the ERTS-1 Satellite, (Source: NASA)

6.3 Data collection systems for transmissions to satellites

In modern space surveys a data collection system (DCS) provides the capability to collect, transmit, and disseminate data from Earth based sensors. Normally such a system involves data collection platforms, satellite relay equipment, ground receiving site equipment and a ground data handling system (Fig. 6.2). The data collection platform (DCP) collects, encodes and transmits ground sensor data to the space platform (e.g. ERTS/ Landsat observatory). The general characteristics of the data collection platform are shown in Fig. 6.3a and b. Such a platform will accept analog, serial-digital, or parallel-digital input data as well as combinations of those. Eight analog bits or 64 bits of digital input can be accepted. The format of a DCP message prior to ecoding consists of 95 bits in the format of Table 6.1.

In the case of the ERTS system the spacecraft acts as a simple relay unit. It receives, translates the frequency and then retransmits the burst messages from the DCPs. No onboard recording, processing or decoding of the data is performed in the ERTS system. In the design of the Spacelab, however, it is anticipated that some on board processing of DCP data from Earth sensors may be carried out.

When the DCP data is retransmitted from the spacecraft it is put on a subcarrier of the Unified S-band (USB) which allows for narrow band telemetry to three primary receiving sites (Fig. 6.3c).

Similar DCP systems are under development for other satellites e.g. the Meteosat. In the Meteosat programme data transmission is planned from land based, ship or buoy data collection platforms. It is anticipated that the Meteosat system will provide 18 channels with a bit rate of 75 bps per channel. Messages can range from 48–4096 bits

Table 6.1
Data collection platform message.

Bits		
1–15	Preamble	
16–17	Synchronization	
18–27	Platform identification	
28–35	Data word number	1
36–43		2
44–51		3
52–59		4
60–67		5
68–75		6
76–83		7
84–91		8
92–95	Encoder run out bits.	

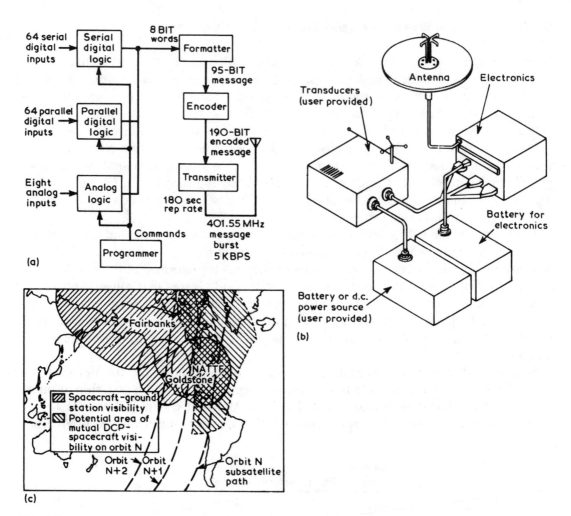

Fig. 6.3

The characteristics of the ERTS-1 Data Collection System. (a) Data collection platform block diagram. (b) Data collection platform. (c) Mutual DCP-receiving site visibility. (Source: NASA)

and an average length of message may be 300 bits. This system has been given successful trials in the Nimbus research and development series.

Looking to the future one may envisage an Earth monitoring system of the kind illustrated in Fig. 17.3 where detection of change in the environmental conditions of the Earth is made possible through automatic sensing systems operating from satellites. Detailed mapping of the conditions on the surface will, however, require a well developed data collection platform system to provide for the calibration of sensors and to gather data that will assist in the analysis of satellite information.

References

Beckett, P.H.T. (1974), 'The statistical assessment of resource surveys by remote sensors, in *Environmental Remote Sensing; applications and achievements,* Barrett, E.C. and Curtis, L.F. (eds). Edward Arnold, London, pp. 11-27.

Curtis, L.F. (1973), 'The application of photography to soil mapping from the air' in *Photographic Techniques in Scientific Research,* Vol. 1. Cruise, J. and Newman, A.A. (eds.), pp. 57-110.

Curtis, L.F. (1974), 'Remote sensing for environmental planning surveys', in *Environmental Remote Sensing,* Barrett, E.C. and Curtis, L.F. (eds.), Edward Arnold, London, pp. 88-109.

Mitchell, C.W. (1973), *Terrain Evaluation,* Longmans, London.

Manual data interpretation and data preprocessing

7.1 Introduction

Although steps are being taken to reduce the dependence of remote sensing data interpretation on the human analyst, data interpretation by manual processes is still of fundamental importance and will remain so through the foreseeable future.

Following training and education an experienced and skilled interpreter can extract much valuable information from such data simply by viewing the image. Sometimes it is sufficient to analyse the image qualitatively in terms of point, line or area features it contains. In such circumstances the task is one of interpretation. Where measurements and map making of an accurate kind are required it is necessary to apply photogrammetric techniques. Whichever objective is in view it is often necessary to pre-process the initial data before photo-interpretation or photogrammetry can be carried out. In the following section we shall deal with the elements of photo-interpretation, photogrammetry and data preprocessing. The latter is reviewed in relation to new techniques designed to improve the data gathered by modern remote sensing systems.

7.2 Elements of photo-interpretation

Photointerpretation has been defined as the act of examining photographic images for the purpose of identifying objects and judging their significance. Although we all 'interpret' photographs reproduced in the media day by day, special training is required for aerial photointerpretation, partly because of the unfamiliar viewpoint of the imagery, and partly because of the special types of information which are usually demanded as the end products of their analysis. Nine elements of photographic interpretation are regarded as being of general significance, largely irrespective of the precise nature of the imagery and the features it portrays. Selected examples are given in the list that follows.
 (a) Shape. Numerous components of the environment can be identified with reasonable certainty merely by their shapes or forms. This is true of both natural features (e.g. geologic structures) and man-made objects (e.g. different types of industrial plant).
 (b) Size. In many cases the lengths, breadths, heights, areas, and/or volumes of imaged objects are significant, whether these are surface features (e.g. different tree species) or atmospheric phenomena (e.g. cumulus vs. cumulonimbus clouds). The approximate scale of many objects can be judged by comparisons with familiar features (e.g. roads) in the scene.

(c) Tone. We have seen how different objects emit or reflect different wavelengths and intensities of radiant energy. Such differences may be recorded as variations of picture tone, colour, or density. They permit the discrimination of many spatial variables, for example on land (different crop types) or at sea (water bodies of contrasting depths or temperatures). The terms light, medium and dark are used to describe variations in tone.

(d) Shadow. Hidden profiles may be revealed in silhouette (e.g. the shapes of buildings or the forms of field boundaries). Shadows are especially useful in geomorphological studies where micro-relief features may be easier to detect under conditions of low-angled solar illumination than when the sun is high in the sky. Unfortunately, deep shadows in areas of complex detail may obscure significant features, for example the volume and distribution of traffic in a city street.

(e) Pattern. Repetitive arrangements of both natural and cultural features are quite common, which is fortunate because much photointerpretation is aimed at the mapping and analysis of relatively complex features, rather than the more basic units of which they may be comprised. Such features include agricultural complexes (e.g. farms and orchards), and terrain features (e.g. alluvial river valleys and coastal plains).

(f) Texture. This is an important photographic characteristic closely associated with tone in the sense that it is a quality which permits two areas of the same overall tone to be differentiated on the basis of microtonal patterns. Common photographic textures include smooth, rippled, mottled, lineated and irregular. Unfortunately, texture analysis tends to be rather subjective since different interpreters may use the same terms in slightly different ways. Texture is rarely the only criterion of identification or correlation employed in interpretational procedures. More often it is invoked as the basis for a subdivision of categories already established using more fundamental criteria. For example two rock units may have the same tone, but different textures.

(g) Site. At an advanced stage in a photointerpretation procedure the location of objects with respect to terrain features or other objects may be helpful in refining the identification and classification of certain picture contents. For example, some tree species are found more commonly in one topographic situation than in others, whilst in industrial areas the association of several clustered, identifiable structures may help us determine the precise nature of the local enterprise. For example, the combination of one or two tall chimneys, a large central building, conveyors, cooling towers, and solid fuel piles point to a correct identification of an installation as a thermal power station.

(h) Resolution. More than most other picture characteristics, resolution depends upon aspects of the remote sensing system itself, including its nature, design and performance, as well as the ambient conditions during the sensing programme, and subsequent processing of the acquired data. Resolution always limits the size and therefore in many cases, the nature, of features which might be recognized. Some objects will always be too small to be resolved (e.g. fair weather cumulus clouds on the average low-orbiting weather satellite photograph), while others lack sharpness or clarity of outline (e.g. the exact position of a shoreline is often difficult to deduce from an air photo of average scale, say 1:10 000).

(i) Stereoscopic appearance. When the same feature is photographed from two different positions with overlap between successive images, an apparently solid model of the feature can be seen under a stereoscope (Fig. 7.1). Such a model is termed a stereo model and the three dimensional view it provides can aid interpretation. This valuable information cannot be obtained from a single print and is usually not available from scanner images. Further discussion of stereoscopic images can be found on p. 114.

In practice these nine elements assume a variety of ranks of importance. Consequently, the order in which they may be examined varies from one type of imagery to another, and from one type of study to another. Sometimes they can lead to assessments of conditions not directly visible in the images, in addition to the identification of features or conditions which are explicitly revealed. The process by which related invisible conditions are established by inference is termed 'convergence of evidence'. It is useful, for example, in assessing the profitability of an enterprise, say a marginal farm, the social class and/or income group occupying a particular neighbourhood, or the soil moisture conditions in agricultural areas.

Photo-interpretation may be very general in its approach and objectives, say in the case of terrain evaluation or land classification. On other occasions it is highly specific, related to clear-cut goals in such fields as geology, forestry, transport studies and soil erosion mapping. Fig. 7.2 is a sample photo-interpretation key for land use and vegetation mapping. In no instance, should the interpreter fail to take account of features other than those for which he is specifically searching. Failure to give adequate consideration to all aspects of a terrain is, perhaps, the commonest source of interpretation error.

The interpretation of aerial photographs is, therefore, an essentially deductive process, and the identification of certain key features leads on to the recognition of others. Once a suitable starting point has been selected, the elements listed earlier are considered either consciously or subconsciously. The completeness and accuracy of the results depend upon an interpreters' ability to integrate such elements in the most appropriate way to achieve the objects which he has been set.

7.3 Enhancement techniques

Where the data used for photo-interpretation consists of multispectral photography or scanning images it is possible to employ enhancement techniques which aim to emphasize the tonal differences between objects. For example, three frames may be selected from a nine frame set because they show the most marked differences. These frames N1, N6, N7 can then be treated as shown below. (Table 7.1).

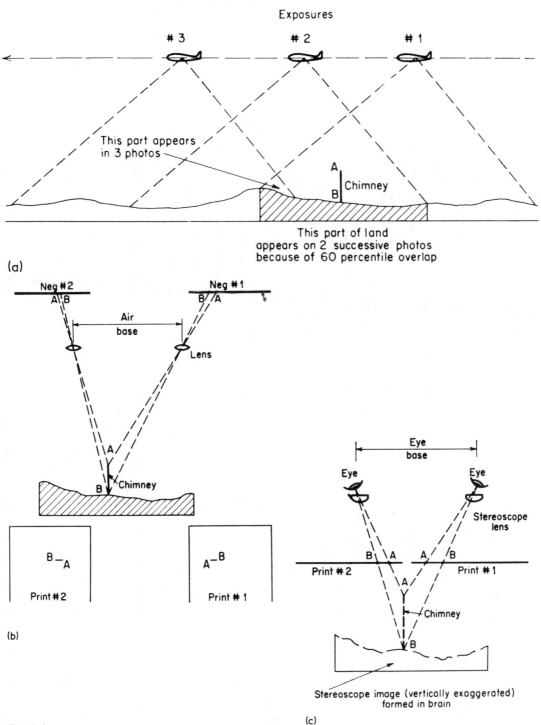

Exposures

#3   #2   #1

This part appears
in 3 photos

A
Chimney
B

This part of land
appears on 2 successive photos
because of 60 percentile overlap

(a)

Neg #2
A B
Neg #1
B A

Air
base

Lens

A
B Chimney

B—A
Print #2

A—B
Print #1

(b)

Eye
base

Eye   Eye

Stereoscope
lens

B A   A B
Print #2   Print #1

A
Chimney
B

Stereoscope image (vertically exaggerated)
formed in brain

(c)

Fig. 7.1
*Three schematic diagrams of the process of stereophotography and stereo viewing. (a) overlapping
photography, (b) the registration of a vertical object on negative and print. Note that the chimney
appears to lean on the prints due to height displacement. (c) shows the stereo image as formed in
the brain by viewing the prints.* (Source: Hunting Surveys and Lueder, 1959)

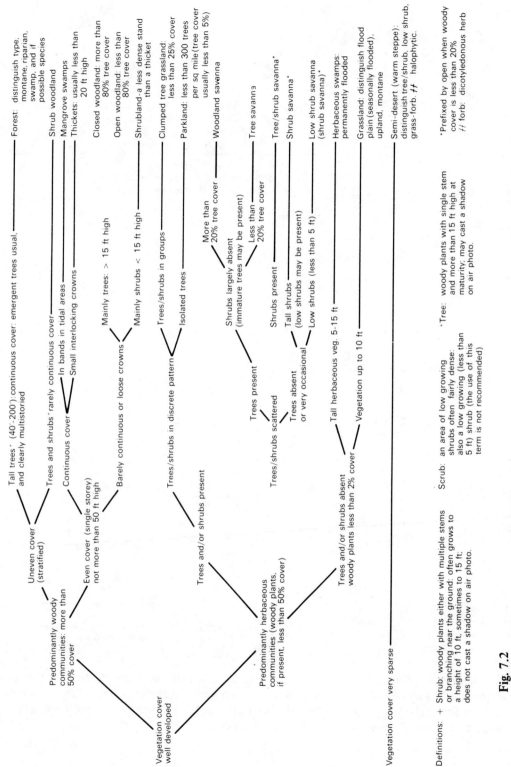

Fig. 7.2

A land use/vegetation key for mapping from aerial photographs in north-east Nigeria. (Source: Alford *et al.*, 1974)

Table 7.1
Enhancement processing.

Camera negatives	Film positives	Duplicate negatives	Intermediates	Additive printing filters	Integral colour print
N1 ⟶	P1 ⟶	N1′	I_1 (N1′ + P6) ⟶ G ⟶		Colour
N6 ⟶	P6		I_2 (N7′ + P1) ⟶ R ⟶		derivative
N7 ⟶	P7 ⟶	N7′	I_3 (N1′ + P7) ⟶ B ⟶		

The positives and negatives are printed subtractively (superimposed) in registration on to pre-punched film to form intermediates. Each intermediate is made from super-imposition of one positive and one negative each from a different band. Each intermediate is then printed additively with a filter so that the image of each one is trans-ferred to only one of the colour layers in the colour film (i.e. I_1 = magenta layer; I_2 = yellow layer; I_3 = cyan layer). The resulting images when developed from a single tri-pack film provide the colour derivative transparency for viewing. The intermediates represent differences of reflectance between bands. Where no differences occur, the negatives and positives cancel out each other, so that no colour shows on the final enhancement. It has been found that the relative brightness and colour differences of objects appearing in the colour derivatives can be used to detect such changes as moisture content, soil density and vegetation conditions.

From the above description it will be evident that the making of colour enhancement prints is complex and time-consuming. It is, therefore, somewhat expensive and it is not always easy to ascertain the best selection of negatives which will give the maximum amount of information at the interpretation phase. In these circumstances, it is an advantage to have the capability of experimenting with different combinations of the negative frames by means of a visual display. In order to achieve this a method is required for projecting spectral positive transparencies, one superimposed upon the other in accurate registration while illuminating each with a different coloured light. If the transparencies are illuminated by the primary colours, a full colour reproduction of the scene is produced. It is also possible to make reconditioned false colour (infrared colour) and artificial derivatives of various kinds by suitable combinations of trans-parencies and colour addition. Additive colour viewers (Fig. 7.3) are important aids to to photointerpretation. The brightness and saturation controls of each spectral trans-parency permit alterations of the final composite screen presentations. In this way subtle differences in the scene can be detected which would not be readily distinguished by examination of normal panchromatic or colour prints.

It is important to bear in mind that the degree of accuracy achieved in photointer-pretation may vary considerably depending on the nature of the subject, the type of photography and the skill of the interpreter. Consideration must be given, therefore,

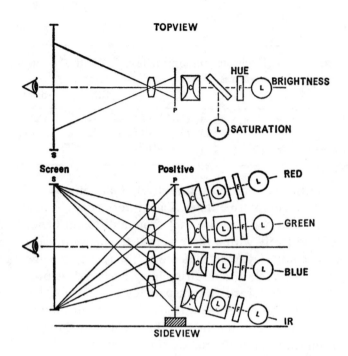

Fig. 7.3
*Additive colour viewing. The positive transparencies (P) are illuminated by light from lamps (L)
passing through filters (F), inner saturation lamps controlling colour saturation and condensing
lenses (C). They are then viewed on the screen (S). (Source: Curtis, 1973).*

to the question of how much ground checking will be required to produce a result comparable in objective accuracy to a ground survey.

Whereas a map offers classification and symbolization of the features it records the
photograph does not. This lack of symbolization makes a photograph more difficult
to interpret but it also makes it potentially more flexible and informative for those
prepared to set up their own classifications on the basis of the image characteristics.

7.4 Measurement and plotting techniques: selected aspects of photogrammetry

Various aspects of plotting from scanner data and scanner images are discussed elsewhere in relation to particular types of image data (e.g. radar, infrared linescan). This
section will, therefore, concentrate on measurement and plotting from aerial photographs. The science of photogrammetry deals with the techniques involved and a
detailed consideration of photogrammetric techniques cannot be given in the space of
this volume. The reader is, therefore, directed towards further reading (see end of
chapter) and selected points of background knowledge are presented here.

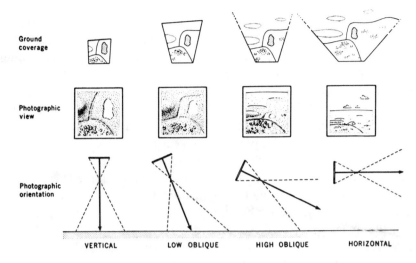

Fig. 7.4

The characteristics of vertical and oblique photographs taken from the air. The high oblique view is similar to that obtained from high ground or a tall building. The horizontal view is like that obtained from a slight eminence. (Source: Spurr, 1960)

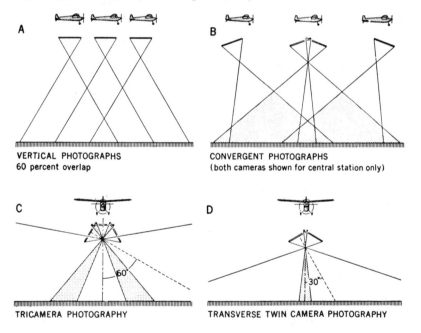

Fig. 7.5

To compare and contrast multiple and single camera aerial photography. The conventional spacing of single camera vertical photography to provide 60 percent overlap along the flight strip (a) is contrasted with tri-camera photography (b) and transverse twin-camera photography (c). (Source: Spurr, 1960)

Aerial photographs may be either vertical or oblique according to whether the camera axis is vertical or not. A 'high oblique' includes the visible horizon whereas a 'low oblique' does not (Fig. 7.4). Oblique photography is sometimes obtained by single cameras but it is more frequently obtained by multiple camera arrangements of the trimetrogon type. In this a single, vertically orientated, camera is flanked on either side by oblique viewing cameras (Fig. 7.5). These systems are cheaper than one employing a single vertical camera, since this requires more flight lines to cover a given area. However the lateral distortions of scale are greater in obliques and less easy to correct.

It is important to recognize that all air photographs are perspective pictures and various types of planimetric inaccuracy are contained within them. These inaccuracies chiefly arise from scale distortions due to height differences in the objects viewed and to tilt distortions (Fig. 7.4). In addition the planimetric accuracy of objects is affected by the displacement of the image due to height variation (Fig. 7.1b).

In order to overcome these problems of scale distortion and height displacement it is necessary to use overlapping photography capable of providing a stereoscopic image. Before considering the techniques employed attention must be focussed on the principal geometric characteristics of airphotographs which are fundamental to the photogrammetric processes. The geometrical properties of the principal point, isocentre and nadir point are particularly important (Fig. 7.6). Briefly stated they are as follows:

(a) A photograph is angle true with respect to its isocentre for all points in the plane of the isocentre.

(b) Straight lines drawn through the photographic nadir point pass through features which would lie on a surveyed straight line on the ground.

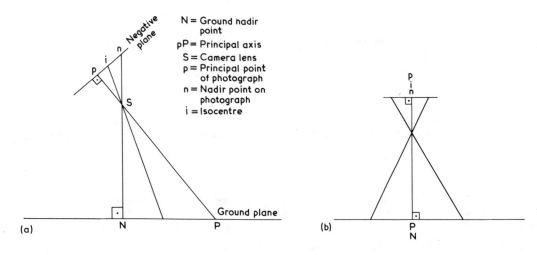

N = Ground nadir point
pP = Principal axis
S = Camera lens
p = Principal point of photograph
n = Nadir point on photograph
i = Isocentre

Fig. 7.6
The geometric characteristics of photographs. (a) Tilted photograph to show principal point, isocentre and nadir point. (b) Vertical case when principal point, isocentre and nadir point coincide.

(c) When a photograph is vertical the isocentre and nadir point are coincident with the principal point which then assumes their properties.

In these circumstances angles measured from the principal point are as truly representative of angles between features as if they had been measured with a theodolite on the ground. Straight lines measured from the principal point also cross detail which would lie on a surveyed straight line on the ground.

These properties are only completely true when the camera axis is perfectly vertical. However, providing the tilt of the camera does not exceed 2^0 and the height variation is small in relation to the flying height (less than 10 per cent) the properties can be held to apply.

Thus is is possible to use the principal points of a number of photos as though they were plane table stations. Rays can be drawn from the principal point on each photograph through points of detail. Intersections can then be made by tracing off the rays from each photograph as they are successively placed under a transparent overlay along a base line. This method is termed the radial line plot and it provides a manual method of plotting the true positions of objects from air photos.

Certain mechanical aids are used in this work by commercial firms. The rays are scribed on transparent film templets and then slots are cut to correspond to the rays. The technique is termed the Slotted Templet method. The rays can also be scribed on a mechanical plotter termed the Radial Line Plotter.

Any method requiring measurement of the height of objects from air photographs is also dependent on stereoscopic cover between successive images. For this reason most vertical photographs are taken overlapping along the flight line (vertical line overlap). In this system features are photographed from successive camera positions (S1, S2 Fig. 7.7). Surface features (e.g. A, B, Fig. 7.7) will occupy different positions

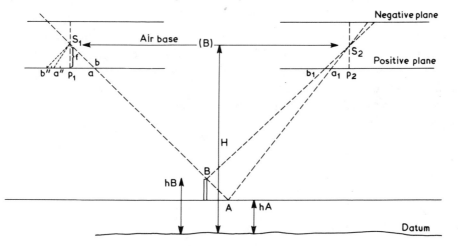

Fig. 7.7
Diagram to show the parallax of points A and B on two photographs taken at two positions S_1 and S_2.

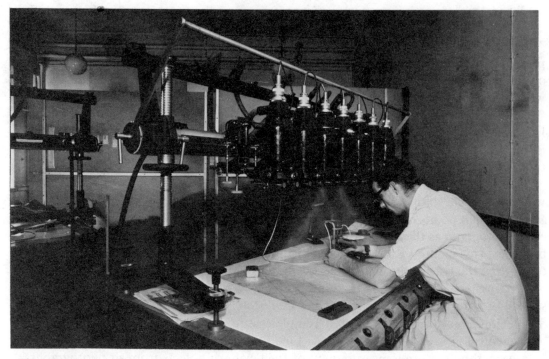

Plate 7.1
A Multiplex plotter in operation. The horizontal bar carries seven projection heads. A stereo model is projected on to the circular white disc which can be raised or lowered to establish the heights of objects. A tracing pencil beneath the white plattern traces the outlines of features to be mapped. (Courtesy Fairey Surveys, Ltd., Maidenhead, Berks.)

in relation to the principal points P1, P2 according to the heights of the features above a given datum. Thus the features A, B are imaged at the same photo point a, b on S1 but at different positions on S2 (a′, b′). If we add the distances P1a and P2a′ on the two positives this gives us a measure of the parallax of A. Similarly if we add the distances P1b and P2b′ we can obtain a measure of the parallax of B. The difference between these two parallax measurements arises from the height difference between A and B, therefore differences in parallax can be used to calculate height differences from air photos.

It is perfectly feasible to measure parallax on photographic prints with a ruler but generally speaking the differences are small so that a more accurate method is demanded. Therefore the Parallax Bar (or stereometer) has been developed which allows small differences in spacing to be measured with a micrometer gauge.

For contouring work it is necessary to employ optical-mechanical devices which will allow an operator to view the stereomodel (Plate 7.1) together with movable markers which can be adjusted in turn to positions A or B. When viewed stereoscopically such markers appear as a single image placed directly on the feature in the stereomodel. If the markers are moved closer together (i.e. parallax increased) the 'fused' mark will appear to float above the feature in the streomodel. If they are separated (i.e. parallax reduced) the marks normally fail to be held in fusion by the brain and they separate on the model.

A skilled operator can manipulate a 'floating mark' of this kind so that it rises and falls on the model, just resting on the surface all the time. By doing so an automatic trace can be made which shows the true plan position of the features traversed. Alternatively the operator can place the floating mark at a known height and then move it across the stereomodel following the contours of the surface. By this means a contour map of the stereo image can be made.

Various types of 'floating' mark equipment are used in photogrammetry. Some instruments depend on the principle of anaglyph projection (i.e. one image red, the other blue, viewed through spectacles with one red and one blue lens). An example of such equipment is the multiplex projector. In this system the floating mark normally consists of a spot of light in the centre of a white disc on a movable tracing table. (Plate 7.1).

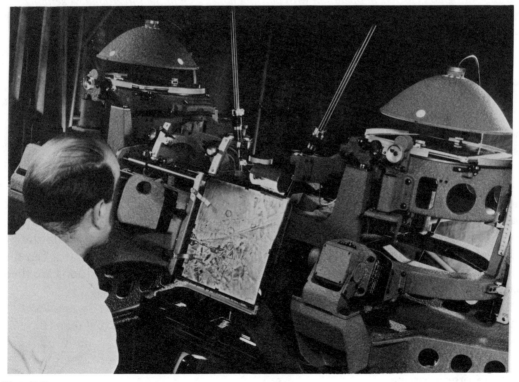

Plate 7.2
A Wild A.8. Stereo Plotter in which diapositives are viewed stereoscopically. The diapositives are mounted beneath the illumination lamps on either side of the apparatus. Note the 'space rods' projecting above the centre of the plotter. (Courtesy Fairey Surveys, Ltd., Maidenhead, Berks.)

Plate 7.3
Wild A.5 Autograph plotters in operation with plotting tables seen in the foreground. A scribing pencil is automatically controlled as the operator manipulates the stereoplotter, placing the floating mark over the points of detail to be mapped. (Courtesy Fairey Surveys, Ltd., Maidenhead, Berks.)

Larger and more expensive manually operated instruments are used for precision work. Many of these have 'space rods' incorporated into the design which accurately reconstruct the angular relationships within the stereo model by mechanical rather than optical means. (Plate 7.2). In this equipment the operator manipulates handwheels and foot controls in order to move and adjust the floating mark on the stereomodel. As he does so the scribing pencil draws on the table beside the stereoplotter.

Very high accuracies can be achieved by high order plotting machines. The limitations in plotting accuracy are set by the nature of the stereomodel produced by the photography and by the ground control available for setting the model into its correct spatial orientation. Thus it should be borne in mind that accurate photogrammetric work also requires accurate ground control. This is normally achieved by establishing the heights of selected well defined features on the photographs. The operator can then orientate the stereomodel in relation to a true datum and heights or positions are then maintained in true relationships within the model.

These manual techniques of photo interpretation and photogrammetry are widely employed by commercial air survey companies and many of our existing maps have been constructed by these means. (Plate 7.3).

7.5 Data preprocessing

7.5.1 General considerations
Most of the data recorded by remote sensing apparatus are not immediately suit-

able for further processing by manual or automatic interpretation. In manual data interpretation there are broadly two types of pictures which may be studied — photographs and images. The former are obtained directly by cameras and the photographic process whereas the images are recorded by indirect means, usually by scanning devices producing an analog signal which is then converted into an image on a screen for reproduction purposes. It will be understood that the data first recorded by the remote sensing apparatus, whether it be the latent photographic image or the sensor signal, cannot be used for interpretation or photogrammetric purposes until it is transformed into a different format. The initial record made by the sensor may consist of one of the following:

(a) Analog voltages recorded on magnetic tape.

(b) Analog voltages converted into digital form and recorded on magnetic tape or punched paper tape.

(c) Latent photographic images.

In order to proceed with further work the data usually has to be converted into one of the following formats:

(a) Photographic transparencies or positive prints.

(b) Computer-compatible digital magnetic tapes.

(c) Computer print-outs, plots, diagrams.

(d) Analog magnetic tapes.

It is possible, therefore, to identify three main tasks at the beginning of the data handling process. These can be summarized as follows:

(a) Processing of photographic material.

(b) Analog-to-digital conversion of signals recorded on magnetic tape and production of computer-compatible tape (CCT).

(c) Digitization of images and, conversely, production of images stored in digital form on magnetic tape.

7.5.2 Photographic processing

Photographic processing can consist of either black and white technology or colour technology. In black and white processing attention has to be given to photographic variables such as characteristic curve, spectral sensitivity, modulation transfer function, dimensional stability, granularity, silver reduction and printing paper. The characteristic curve shows density as a function of log exposure. It can be seen (Fig. 7.8a) that density is nearly proportional to the log of exposure in the central part of the S-shaped curve. The proportionality factor as measured by the ratio a/b is referred to as the film 'γ'. Where density measurements are to be made on a number of films it is necessary to process each film to the same γ. Spectral sensitivity of the film describes the sensitivity of the film in a given region of the spectrum. Panchromatic films are sensitive up to 700 nm and must be developed in complete darkness whereas certain duplicating films are sensitive to 500 nm and can be developed in red light. The modulation transfer

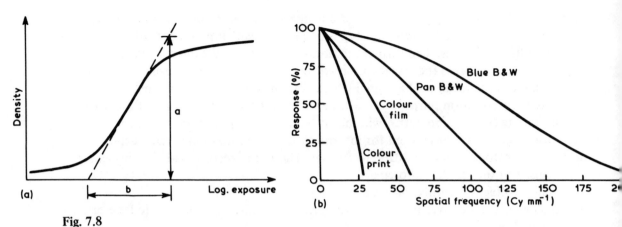

Fig. 7.8
(a) Shape of typical characteristic curve for photographic film. (b) Film transfer function curves.

function (MTF) is the ratio of intensity variations in the image to those occurring in the original. It is an expression of the resolving power of the film and the achievement of maximum resolution depends on the chemistry and duration of the photographic process. The dimensional stability of film used in the photographic process is important in order to limit distortion in the image.

Stability improves with thickness of the film but in space applications films must be as thin as possible in order to conserve weight (see pp. 54). It is also necessary to control the film environment if film stability is to be maintained — especially the temperature and humidity. Polyester base material is usually superior to other materials in dimensional stability. For example its expansion coefficient is normally 0.001−0.01 per cent per $^\circ$F whereas cellulose triacetate is less good by a factor of 2−3 times. The granularity of a film becomes important because it represents unevenness in the emulsion which creates 'noise' in the photograph image. A microdensitometer trace on a uniformly exposed and processed film can be used to detect irregularities or discontinuities and thus assess its granularity. Root mean square graininess can be determined from the standard deviation of the density measurements and from the diameter of the scanning aperture:

$$G = K\sigma_K (D) \tag{7.1}$$

where
G = Rms graininess
K = diameter of the scanning aperture used
$\sigma_K(D)$ = standard deviation of density measurements.

The signal to noise ration (S/N) can be calculated using the following equation:

$$S/N = \frac{\Delta D}{\sigma(D)} = \frac{\Delta D \, (A)^{\frac{1}{2}}}{G} \tag{7.2}$$

where $\qquad$ G = Rms graininess.

ΔD = incremental density change

G = area of aperture.

In processing photographic film it is usually necessary to control the density of the product. This is done by a method which removes silver from the negative and is termed reduction. Ammonia and potassium persulphates are reducers which are commonly used but their action is liable to be uneven and hard to control. Other methods include 'dodging' which may be carried out manually by inserting tissue paper or a mask to decrease the density range. Automatic dodging printers are available and are reliable and efficient. The printing papers used in the photographic process can accommodate a wide range of density ranges and chloride, chlorobromide and bromide papers are the major types used.

In colour technology the chief areas of interest are in dyeing and reversal processes, resolution, colorimetry, colour balance and reproducibility.

The colour films used are primarily natural colour and infrared colour. The use of aerial colour only became significant in the 1960's following the breakthrough achieved by film manufacturers who succeeded in producing a range of colour emulsions with speed and resolution characteristics comparable to panchromatic films. Aerial colour films are available as either reversal colour films or negative colour films. Generally, the reversal films are used for medium to high altitude photography and negative colour films for work at low altitudes.

It has been appreciated for some time that the visual contrast presented by colour to the human eye is much greater than that of panchromatic photography. The eye will distinguish about 200 gradations on a neutral or grey scale, whereas it is capable of differentiating some 20 000 different spectral hues and chromas (intensities of colour). Therefore colour photography immediately offers the opportunity of more detailed study of surface objects.

Infrared colour is a false-colour reversal film. It differs from ordinary colour film in that the three sensitized layers are sensitive to green, red and infrared radiation instead of having the usual blue, green and red sensitivities. The green sensitive layer is developed to a yellow positive image, the red sensitive to a magenta image and the infrared to a cyan positive image, thus the colours are false for most natural objects.

False colour film is highly sensitive to the green-red wavelengths of light as well as the near infrared. As a result it has particular characteristics such as that water and wet surfaces image in blue and blue grey tones. An important feature of this type of film is that it records healthy green vegetation in various shades of red in the positive images (see Tables 4.1 and 4.2).

Colour films are generally characterized by lower spatial resolution and higher contrast than black and white (Fig. 7.8b). Colour balance is particularly important in treatment of false colour films because unlike in true colour photography the user has no reference available for comparison. The main aim of colour balancing is to ensure that brightness differences occurring in the target area are reproduced faithfully in the film image without affecting hues and colour response. In order to achieve this adjustments of the exposure for each layer are required. Sequential printers are available which allow independent layer corrections. Problems sometimes arise, however, due to variations in the characteristics of stored false colour film. These can be largely overcome by storing the film at -20° C and making some densitometric tests before processing.

The photographic equipment necessary for photographic processing includes enlargers, contact printers, film and paper print processors, mixers, drier/cutters, microfilm processors and light tables.

7.5.3 Optical and digital conversion

A fundamental requirement in Earth sensing studies is the ability to convert images between the optical and digital domains. Computers can interact with optical data by means of optical-digital (O/D) converters and conversely D/O converters. If this facility is available it becomes possible to store a picture in digital form or to produce a picture from digital data. Normally some form of scanning device is used in the process of either 'reading' or 'writing' the image. The principal types of equipment used are (a) drum microdensitometer and recorder (b) CRT readers, writers and displays (c) electron beam and laser recorders.

Drum type machines are essentially high resolution microdensitometers where the film is fastened on to a rotary drum and scanned by a light source whilst in rotary motion. Traditional light sources are normally used but exceptionally fast machines have been developed using laser beams. A typical high speed digital microdensitometer is the System P–1000, Photoscan, which provides the following scanning times for a 12.5 x 12.5 cm film (Table 7.2). A schematic diagram showing the basic characteristics of the drum scanning microdensitometer is shown in Fig. 7.9a.

Flying spot CRT and vidicon systems (Fig. 7.9b) offer high speed of conversion, easy interaction with computers, possibility for data reduction by selective access to images for spectral separation, and for filtering and image improvement techniques. These machines can read 64 levels on a grey scale and record 60 levels and require 20 μs for reading and 27 μs or less for recording. Spot sizes of 0.0012–0.002 in can be selected for scanning.

Electron beam recorders (EBR) are image printers using input data which may be in analog form (video) or digital form. The basic EBR consists of a high resolution electron

Table 7.2

Drum speed (rev s^{-1})	Raster (μm)	Data rate (Hz)	Scantime (min)
2	50	14.4	20
4	100	14.4	5
8	200	14.4	1.25

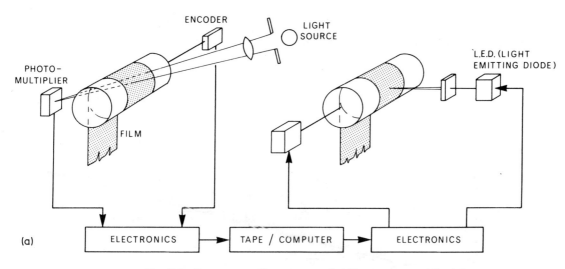

Fig. 7.9 *Image recording systems. (a) Drum scanner principles.*

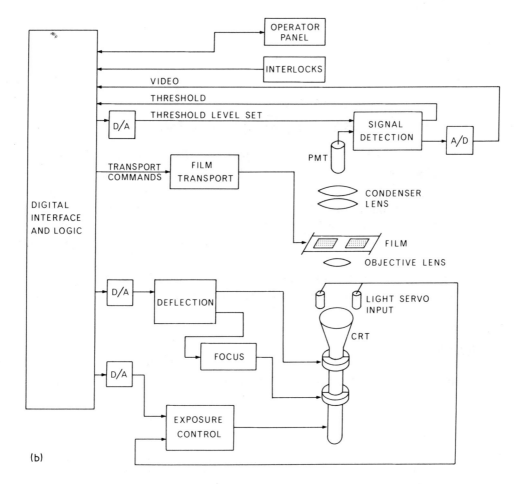

Fig. 7.9 *Image recording systems (b) Flying spot scanner schematic diagram.*

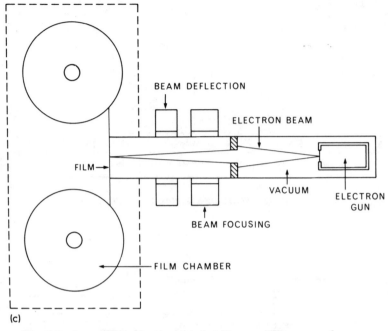

Fig. 7.9 *Image recording systems (c) Electron beam recorder.*

gun, and electron optical system for controlling the electron beam, a film transportation mechanism, an automatic vacuum system and regulators and electronic circuits which operate the recorder. A schematic diagram of an electron beam recorder is given in Fig. 7.9c. The electron gun provides an electron spot (3–10 μm diameter) which is focused by coils on the sides of the vacuum tube.

Laser scanners consist essentially of three components (a) optical system, (b) film transport system and (c) rotating scanner system. The laser provides a high power, collimated beam of monochromatic light which scans the film image. Typical specifications for commercially available equipment are given in Table 7.3.

7.5.4 Analog to digital conversion

The digitizing procedure is accomplished by an A-D converter in a number of steps. The area to be digitized (resolution element) is selected by 'gate' settings which can be set up manually on the A-D converter so that only certain regions are digitised. When the 'gated' section is read by the analog tape recorder the data value of each channel is carried into a system which shifts right the appropriate number of places dictated by a data resolution switch which incorporates the number of significant bits used for each data value.

When the computer word is completely filled a transfer takes place whereby the computer stores the word from the A-D converter in sequential addresses in core. After

Table 7.3
Laser scanner characteristics (RCA, Model LR-70 and LR-71).
(Source: ESRO CR-295, 1973).

Model	Parameter	Performance
LR-70	Resolution (@ 50% MTF)	20 000 pixels per scan
	Video bandwidth	DC to 75 MHz
	Geometric fidelity	1 part in 20 000
	Grey scale	$16 (2)^{1/2}$
	Film format	5 in x 1500 ft
	Scan rate	To 5000 scan s^{-1}
	Film velocity	Variable from 0.25 in s^{-1}
	5 in frame scan time	4 s
LR-71	Resolution (@ 50% MTF)	100 1p mm^{-1}
	Video bandwidth	DC to 30 MHz
	Geometric fidelity	1 part in 20 000
	Grey scale	$16 (2)^{1/2}$
	Film format	5 in x 1500 ft
	Scan rate	Variable to 7500 scans s^{-1}
	Film velocity	Selected ranges between 0.15 to 40 in s^{-1}

the final scan line has been digitized and transferred to the computer a signal is sent to the computer that the block of data is complete. The computer then places the digitized information onto magnetic tape in the form of one scan line per digital record.
Often there is overlap between scan lines made by the sensor i.e. the same part of the ground region is sampled twice. This results in a superfluous amount of analog data (oversampling has occurred). In these circumstances a 'smoothing' or 'filtering' programme can be incorporated in the digital computer. This programme normally has two objectives. First, to reduce the number of scan lines by an integer factor which will give true 'aspect' to the ground sampling. Second, to reduce the instrumental 'noise' present in the recording process. This can be accomplished by averaging of data over a number of lines.

7.5.5 Digital to analog conversion
With the development of multispectral scanning devices it is now necessary to have a means whereby digital data can be converted into analog data. Such a system enables photographic imagery to be generated from digital data.
Generally D-A conversion is achieved by reading digital data in the computer from computer compatible tape, transforming it into a form acceptable for digital-analog hardware equipment, and then passing the transformed data into the D-A device. Since

analog tape recorders are not able to stop and start rapidly (unlike digital recorders) it is desirable to maintain a continuous flow of data to the analog hardware. This is normally achieved by setting up an input and output queue. The system is arranged so that there is always a backlog of input data waiting for processing and also a reserve (buffer) amount of data in the output to the analog hardware. This enables corrections to the input data to be made without interruption of flow to the analog device.

7.5.6 The Landsat System

The Landsat Data Collecting System consists of a combination of ground station, operations control centre and NASA Data Processing Facility (NDPF) as shown in Fig. 7.10.

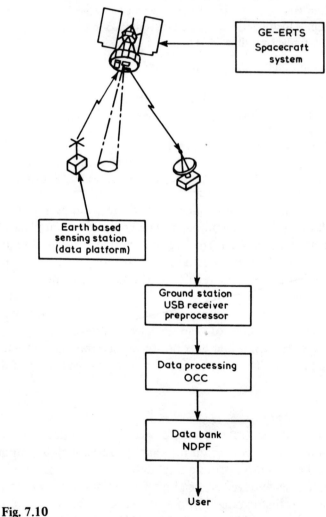

Fig. 7.10
Elements of the ERTS DCS (Data Collection Service) system. (Source: NASA)

The scale of the data processing operation can be assessed by considering production requirements for RBV (Table 7.4) and MSS sensors (Table 7.5).

Two data treatments are available in the image processing systems: (i) photographic processing (ii) analog to digital special processing which leads to the production of computer readable tapes. (Fig. 7.11). Thus investigators using Landsat (ERTS) data can obtain photographic images and computer readable tapes for their studies. These can be purchased from EROS Data Center, Sioux Falls, U.S.A.

Table 7.4

Data processing facility requirements for Landsat return beam vidicon observations.
(Source: NASA II – 19/20, 1972)

RBV input (3 bands)	Scenes per week (%)	Scenes per week	For each scene	Items per week
1316 scenes/wk 3948 images/wk	100% Bulk	1316	3 B–W Masters	3 948
			30 +	39 480
			30 –	39 480
			30 + Prints	39 480
	20% Bulk colour	263	1 C –	263
			10 C Prints	2 630
	5% Precision	66	3 B–W Masters	198
			30 –	1 980
			30 +	1 980
			30 + Prints	1 980
			1 C –	66
			10 C Prints	660
			Ability to digitize	–
	1% digitized	13.2	3 Copies computer readable	$\approx$ 158 tapes

Key: B–W Black and White
   C   Colour
   +   Positive transparency
   –   Negative transparency
   + Prints Positive paper prints
   C Prints Colour positive paper prints

Table 7.5

Data processing facility requirements for Landsat multispectral scanner system observations. (Source: NASA).

MSS input (4 bands)	Scenes per week (%)	Scenes per week	For each scene	Items per week
1316 scenes/wk 5264 images/wk	100% Bulk	1316	4 B–W Masters	5 264
			40 +	52 640
			40 –	52 640
			40 + Prints	52 640
	20% Bulk colour	263	2 C –	526
			20 C Prints	5 260
	5% Precision	66	4 B–W Masters	264
			40 +	2 640
			40 –	2 640
			40 + Prints	2 640
			2 C –	132
			20 C Prints	1 320
			Ability to digitize	–
	5% Computer readable	66	3 Copies computer readable	≈ 713 tapes

Key: B–W   Black and White
      C   Colour
      +   Positive transparency
      –   Negative transparency
      + Prints   Positive paper prints
      C Prints   Colour positive paper prints

When the basic procedures of photo-interpretation and photogrammetry were developing some sixty years ago, the amount and range of image data were limited. The explosive development of remote sensing techniques has necessitated the evolution of a wide range of preprocessing methods. These modern methods are designed to improve the precision of the data to be analysed and to allow surplus data to be discarded.

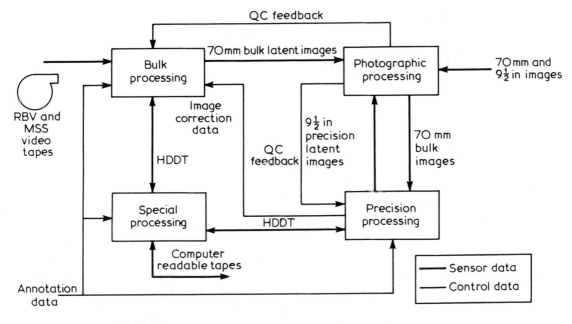

Fig. 7.11

Image processing subsystems for Landsat. (Source: NASA)

References

Alford, M., Tuley, P., Hailstone, E., and Hailstone, J. (1974), 'The measurement and mapping of land-resource, data by point-sampling on aerial photographs, in *Environmental Remote Sensing: applications and achievements,* Barrett, E.C., and Curtis, L.F. (eds.), Edward Arnold, London, pp. 113-126.

American Society of Photogrammetry and Society of Photographic Scientists and Engineers, (1969), *New Horizons in Color Aerial Photography,* Falls Church, Virginia, p. 423.

Bressanin, G. and Erickson, J. (1973), *Data Processing Systems for Earth Resources Surveys,* European Space Research Organisation, Noordwijk.

Cimerman, V.J. and Tomasegovic, Z. (1970), *Atlas of Photogrammetric Instruments,* Elsevier, Amsterdam.

Curtis, L.F. (1973), 'The application of photography to soil mapping from the air', in *Photographic Techniques in Scientific Research, Vol. 1,* (eds.), Cruise, J. and Newman, A.A., Academic Press, New York and London, pp. 57-110.

Lueder, D.R. (1959), *Aerial Photographic Interpretation,* McGraw-Hill, New York.

Spurr, S.H. (1960), *Photogrammetry and Photointerpretation,* 2nd Edn., Ronald Press, New York, pp. 13-36.

Wenderoth, S., Yost, E., Kalia, R. and Anderson, R. (1974), *Multispectral Photography for Earth Resources,* Remote Sensing Information Centre, Greenvale, New York, p. 265.

# Numerical processing and analysis

8.1 The need for numerical processing techniques

The analysis and interpretation of remote sensing data by manual means has much to recommend it. For example, the trained human eye can identify a very wide range of imaged features with both ease and accuracy. Consequently the interpretation of aerial photographs, space photographs, and other forms of remote sensing imagery is, perhaps, best carried out by hand if the data sets are small. But the volume of data generated by new remote sensing systems is growing rapidly, and already far exceeds the capability of trained interpreters in some environmental fields. It has been estimated that a single low-altitude Earth-orbiting satellite such as any of the Noaa (National Oceanic and Atmospheric Administration) weather satellites can yield something of the order of $10^{10} - 10^{12}$ points of new data every day. Members of the ERTS family planned for the future may be capable of even greater outputs, yielding as much as 250 million data bits per second. Still further increases may be expected as bandwidth restrictions are eased. Computer-based interpretation techniques seem to afford the only possible solution to the problem of routine extraction of useful facts on a near-real time basis. Numbers are required for environmental hypotheses to be tested rigorously. Although manual analyses can provide some such information, data arrays of suitable magnitude for statistical processing can usually be compiled much more quickly by automatic analytical techniques.

So the choice at present in the analyses of images for selected objects lies between range of features and accuracy of identification on the one hand, and numbers of cases and speed operation on the other. Where more general picture characteristics are concerned (for example, picture brightness and texture) the eye does not perform well at the task of discriminating differences, and more 'objective' techniques may be preferred. Most of these are capable of providing numerical answers to reasonable questions. Numerical methods are especially useful for the development of probabilistic statements about points, lines or areas. Whether by manual or objective means the complete identification or correct classification of features is often not possible. Quantitative techniques are often preferable to qualitative methods in that they can be used to provide accuracy levels of the determinations.

We may list the principal applications of numerical methods of processing and analysis as follows:

(a) In the presentation of remote sensing data in manageable quantities and acceptable forms;

(b) In the processing of such data to enhance or reveal selected phenomena;

(c) In the analysis and interpretation of selected data; and

(d) In the development of procedures for routine application to selected types of data for different purposes. These procedures include those intended for use in operational identification (classification) procedures.

We may profitably examine in greater detail these four areas of activity.

8.2 The quality and quantity of remote sensing data

Not infrequently the data from remote sensing missions are less than optimal in quality. Each remote sensing system has its own characteristic capability depending on such things as the instrument design, altitude of the platform, and performance of the recording equipment.

Other less predictable, often extraneous, influences may adversely affect the quality of the retrieved data (Figs 8.1a and b, and Table 8.1). Methods commonly referred to as 'preprocessing functions' are necessary to reduce or eliminate such inadequacies. Some such methods are implemented using special techniques with film, or with optical and electronic analog systems. (See Chapter 7). Here we may concentrate our attention upon digital systems.

8.2.1 Geometric corrections

Remote sensing data, commonly obtained either in snapshot or line-scan forms, come in a variety of scales and geometries (Fig. 8.2). Often these must be rectified before they can be compared in detail with existing data. The problem arises, for example, in studies designed to investigate the relative merits of imagery from aircraft or satellite platforms for the identification and mapping of features at a selected level of detail in a particular region. The problem is most intransigent where:

(a) Imaged areas are topographically rough.

(b) The images are obtained from systems with broad fields of view.

(c) The images are obtained from low altitude platforms, where image displacement due to quite modest topographic features may be considerable.

(d) Systems other than framing cameras are used.

Rectification can be performed by optical and electronic means (of which the former is often the better and cheaper), or by digital techniques. Basically the problem involves bringing separate data into congruence, that is to say ensuring that point-to-point correspondences are achieved. These are necessary, within certain limits, if point decisions or point comparisons are intended using statistical pattern recognition techniques. The fast Fourier transform algorithm has been employed quite widely to provide the two-

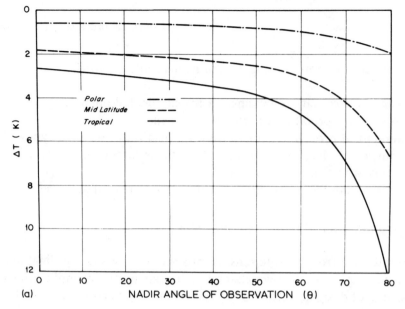

Fig. 8.1

Estimated departures (ΔT) of Nimbus II HRIR equivalent blackbody temperatures as function of local zenith angle for three different air masses. Unfortunately this represents a gross over-generalization of the influence of the Earth's atmosphere on infrared transmissions toward space.

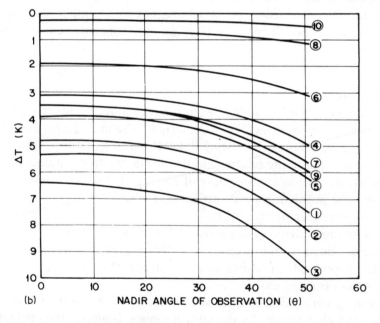

Fig. 8.1

(b) Calculated departures of Nimbus IV THIR 11.5 μm channel temperatures from ground surface temperatures as a function of local zenith angle for ten different atmospheres (See Table 9.1). Unfortunately the structure of the atmosphere is complex and rapidly changing and even these more detailed correction functions leave residual errors in the satellite data after processing. (Source: Sabatini, 1971)

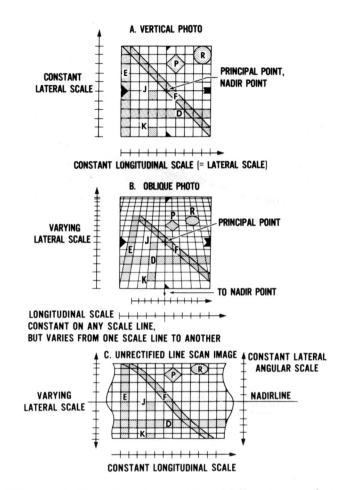

Fig. 8.2
To exemplify the differences in the scales and geometries of different types of remote sensing data.
(Source: Taylor and Stingelin, 1969)

dimensional Fourier transform of an image, and then to facilitate rapid cross-correlation between pairs of images of data arrays.

Since the variations in the performance of an orbiting sensor system are proportionately less than those associated with lower-altitude systems, time-differing images from satellites may be brought into acceptable congruence more easily and more cheaply per unit area than those from aircraft. Herein is one of the principal advantages of a satellite remote sensing system.

8.2.2 Radiometric corrections
These are necessary to lessen the effects of a number of variations in radiation or image density. The aim is to achieve a product with a high level of invariance with respect to

the radiance of the scene, whether the radiance is reflected or emitted energy. Common image defects are associated with:

(a) A decline in light intensity away from the centre of a photographic lens.

(b) The relation between the angle of view and the angle of solar radiation.

(c) Variations in illumination along a sensor flight path. These are especially significant in connection with polar-orbiting satellites.

(d) Variations in viewing angles of sensors scanning across the flight path of the sensor platform.

(e) Image variations arising from the performance of image tubes.

(f) Variations in photo-processing and developing procedures.

Numerical methods of combatting such effects include several 'normalization'

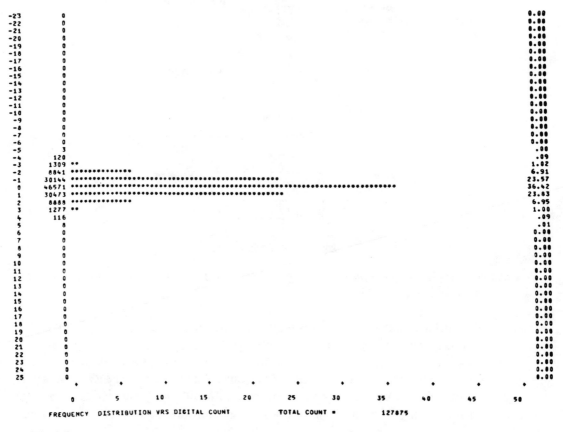

Fig. 8.3
Noise pattern, Noaa 2 tape recorder (absolute response) shown for a supposedly inert recording of night time data from the visual channel scan sensor. This may create systematic streaking in display products, and produce systematic contamination in the generation of more quantitative applications. (Source: Conlan, 1973)

techniques. Most involve the assessment of the nature and degree of the effect caused by one or more of the sources of variation in recorded radiation density, and the development of appropriate physical remedies (e.g. filters or films) or numerical weighting functions (e.g. calibration curves and algorithms) to improve the data.

8.2.3 The reduction of noise

Noise may be defined as any unwanted (either random or periodic) fluctuation of a signal which may obscure it and make its analysis and interpretation more difficult. Random noise may be caused by the performance of remote sensing systems during recording, storing, transmission, and especially in ground reception of the data. Periodic noise may be caused by radio interference, and by certain components of the sensor/platform complexes themselves.

It is possible to dampen random noise in some cases through the application of statistical smoothing functions to the data presented in numerical form, but such noise can rarely be eliminated. Rather, efforts are necessary to keep it within acceptable bounds.

Periodic noise is easier to eliminate because its pattern in both one-dimensional and two-dimensional products is more systematic. Fig. 8.3 illustrates the noise outline for the tape-recorder on Noaa-2. The rotating mirror in the Noaa-2 Scanning Radiometer, and the satellite's tape recorder drive motor, were controlled by a common power source oscillator. Consequently the noise from this source associated with the tape recorder — inherent in the data transmitted to the ground — tended to be periodic. Cloud pictures contained regular 'streaky' patterns, which, if untreated, would have been passed on to secondary products. When the pattern of noise is known, appropriate modifications can be made to ground station hardware to lessen its effect.

8.2.4 Image enhancement

Specific users of remote sensing data often require that the features of special interest to themselves be emphasized or enhanced at the expense of other, background, features. Such enhancement can be carried out photographically (see Chapter 7) or digitally. The procedures include density slicing, edge enhancement (image sharpening), contrast stretching (increasing greyscale differences), and change detection by image addition, subtraction or averaging. The digital approach has the advantage of flexibility, the photographic approach has the advantage of cheapness (see previous chapter). The digital approaches may be summarized as follows:

(a) Density slicing. With a digital system any level or levels of grey may be selected on a photographic or electrical imprint, with output via a line printer, on a flat-bed or drum plotter, or directly onto a photographic film writer. The procedure involves the development of suitable algorithms and computer programmes. Quantization noise or spurious contouring may result if the slices are not chosen carefully.

(b) Edge enhancement. This can be used to sharpen an image by restoring high-frequency components through the removal of scanline noise, or to emphasise certain edges to aid interpretation by taking the first derivative in a given direction. Unfortunately the computer cannot easily distinguish desired lineaments from others of similar greyscale change, and these will be enhanced as well.

(c) Contrast stretching. Any image of low contrast or any portion of the greyscale of an image may be enhanced in digital processing through suitable adjustments to the array of picture points.

(d) Change detection. Change detection is generally carried out by hand, using inter-active man-machine methods, for example the flicker procedure whereby the first and second images are presented alternatively to the observer many times a second. Apart from simple techniques like the detection of greyscale change from one photograph to another, digital procedures have not yet been developed with notable success in this interesting, important, but rather intransigent field.

8.2.5 Calibration

Substantial sections of this volume are concerned with the analysis and interpretation of remote sensing data in terms of familiar characteristics of the human environment. Since our funds of environmental data from conventional sources involving *in situ* sensors are much greater than those of remote sensing data, having been built up over a much longer period of time, it is common practice to calibrate the remote sensing data in terms of conventional parameters. Whether or not this practice will continue after the imbalance in the two data types has been reduced, redressed or even reversed remains to be seen. Already some environmental patterns mapped from satellite altitudes have been expressed simply in terms of the measured radiances (e.g. stratospheric radiances in terms of voltages), with no attempt to calibrate the results in terms of commonly accepted environmental units. At present this approach is restricted mainly to situations in which conventional ('ground truth') data are few or unavailable.

Often prelaunch calibration procedures are carried out to provide the numerical voltage-to-temperature or voltage-to-brightness response relations for the sensor systems. For example, the calibration programme for operational weather satellites involves the preparation of families of calibration tables for each sensor for converting normalized raw counts into effective blackbody radiative temperature responses in the case of infrared sensors, or into calibrated brightness response for the visual channel. These values are then corrected for any bias as a function of sensor temperature, a measurement regularly available in the telemetry data.

The final step involves those corrections necessary for the interpretation of mea-surements of the natural Earth scene. For example, allowances must be made for atmospheric attenuation in the case of infrared sensors. This 'limb darkening' cor-rection is generally taken as a function of local zenith angle, with the assumption that water vapour is the primary absorbing constituent (Table 8.1).

For the visual channel on current Noaa satellites the final brightness response is

Table 8.1

Computations of atmospheric effects on temperatures derived from the THIR radiometer (11.5 μm channel) on the weather satellite, Nimbus IV. Though detailed, the table represents a broad generalization of a medium capable of large and rapid variations in temperature and moisture content.

Atmosphere	Temperature surface (K)	Precipitable water (cm)	'Observed' equivalent blackbody temperatures at nadir angles of:			
			0°	15°	30°	50°
(1) Standard	288.15	1.96	283.38	283.24	282.77	280.70
(2) Tropical	299.65	3.63	294.37	294.23	293.74	291.56
(3) Subtropical summer	301.15	3.94	294.77	294.60	294.02	291.45
(4) Subtropical winter	287.15	1.93	284.02	283.93	283.61	282.18
(5) Mid-latitude summer	294.15	2.64	290.29	290.19	289.79	287.95
(6) Mid-latitude winter	272.59	0.74	270.64	270.58	270.41	269.53
(7) Sub-arctic summer	287.15	1.84	283.59	283.49	283.13	281.55
(8) Sub-arctic winter (cold)	257.28	0.36	256.58	256.56	256.48	256.13
(9) Arctic summer	278.15	1.65	274.51	274.39	274.00	272.24
(10) Arctic winter (mean)	249.22	0.18	248.95	248.95	248.91	248.74

made as compensation for differences in solar illumination of the Earth scene. The algorithm in use at present is a simple cosine function of the solar zenith angle.

8.2.6 Data compression

The reduction of the raw data obtained from a remote sensing system into a volume no larger than that required for a particular programme of analysis and interpretation may be effected by one or more of a wide range of standard statistical techniques. These include smoothing and averaging processes, sampling techniques, and feature extraction methods. Smoothing or averaging processes are commonly applied by numerical means to data in either one-dimensional (e.g. line scan) or two-dimensional (data array) forms. Supplementary processing may be carried out by the computer to ensure that the assumptions of subsequent numerical processes might be met. For example weighting factors may be applied to normalize the frequency distribution of the new, reduced population. If the original data are to be reduced by selection instead, the principles and practices of statistical sampling theory are invoked. The problems of feature extraction are not difficult to resolve providing that the features to be removed for further study are related to simple characteristics of the raw data themselves, for example regions of radiation temperatures above a given threshold in infrared studies, or areas with selected ranges of brightness in densitometric analyses of conventional photographs. Further data-reduction processes may then be applied to the information selected for careful scrutiny. However, the problems of selective feature extraction are very difficult when combinations of data characteristics are invoked simultaneously. Some of the attendant difficulties of pattern analysis and pattern recognition are discussed in more detail below.

8.3 The analysis and interpretation of selected features

Quantitative feature extraction from remote sensing data involves four steps, namely:
(a) The development of methods whereby significant features or variables may be isolated.
(b) The preparation of appropriate quantitative data arrays.
(c) The determination of the essence of the structure of the data.
(d) The reduction of the information so that only the essential point in, or structure of, the data is carried forward into the final stage of decision classification.

The first step involves the logical or syntactical structuring of the design of the experiment. This is the planning stage in which consideration is given to what will be measured and why; how the data will be sampled and aggregated; how many variables should be used; which methods of data retrieval, preprocessing, and data compaction will be employed; and how the final analytical stages will be organized.

The second step involves some form of line-by-line data scanning to quantize the variables to be used in the analysis. These variables may be uncorrected greyscale (from a photograph, colour waveband or multispectral channel) or a normalized variable. In most remote sensing applications the principal variables used in each channel will be tone, texture or height measurements. These relate to greyscale, greyscale organisation and variability, and stereoscopic characteristics respectively. Resolution on the ground, and resolution in density, or greyscale (the number of detectable shades of grey), may limit the subsequent stages. On other occasions limitations are imposed by the analyst himself, for example with the number of final categories required in his analysis in mind.

The third stage involves the search for key indications of the phenomena or relationships under scrutiny. Often it is sufficient to select a few aerial photographs from a numerous set, or the indications from some rather than all the multispectral channels available. Extraneous or highly cross-correlated data can be eliminated, reducing the measurements or variables to the minimum which is substantially orthogonal in feature space. Thus the fourth step of the extraction process is reached.

In many programmes of data analysis and interpretation attention is focussed either upon point features, or non-point features having length, or length and breadth. In order that image contents may be correctly interpreted and classified, it is common practice for 'training sets' of data to be compiled. These are subsamples whose identification is completely known. They can be used as stereotypes of points, lines or areas, by comparison with which unknown areas can be assessed. In other words, new data may be interpreted through their resemblance to these sets of templates or 'fingerprints' in a 'supervised' approach. The statistical parameters of each training set are calculated by analog or digital computer and involve many factors which may have a bearing on the characteristics of the key site, ground truth station, or ground survey area. These include such factors as topography, vegetation and soil content, farming practices and environmental management as well as time of day, season of the year, and prevalent weather conditions. In calculating the statistical parameters of the training set, the greater the number of characteristics considered at a point, or the greater the number of resolution elements within an area, the better the classification accuracy will be.

Training sets are commonly used where multispectral scanner data are available. The data-display techniques which are used to facilitate feature extraction are numerous, and include histograms for each category, and by each channel. Matrices of correlation and covariance may be compiled. The training set data usually involve parametric discriminants using Bayesian probabilities and maximum likelihood functions. These commonly employ thresholding procedures in which items unlike the sample are assigned to a discard class.

Many methods are being used or tested for comparing new or unclassified data with established training sets. These include nonparametric procedures such as linear discriminant functions which are much faster in terms of computational time than the maximum likelihood decision rule, which, in practice often becomes a quadratic rule requiring a large number of multiplications for each decision. Some workers have used composite sequential algorithms based on clustering techniques. Others have developed models based on boundary conditions. Still others have developed 'table look-up procedures', much faster than maximum likelihood functions in practice and nearly as accurate. Operations in the spatial frequency domain have been investigated for areal analysis and classification.

Analog computer techniques can be operated rapidly to provide image comparisons based on brightness levels, areas above threshold brightness levels, and the numbers and shapes of selected picture features. Unfortunately such systems lack the flexibility of digital computers, and each new capacity or operation demands new hard-wired components.

Conventional training-set techniques are, however, not without disadvantages. In particular the ground truth sites or areas must be carefully chosen, presupposing detailed knowledge of the region. Often such knowledge is not sufficiently detailed to ensure the representativeness of the training set, and further problems are caused by variability in time and space within the training area. Consequently efforts are being made to develop techniques which may be described as 'unsupervised', for application to areas for which training sets are unavailable. Such techniques involve clustering algorithms, which may take the form of manual routines whereby selected features are identified by eye and mapped. The technique relies on the fact that the multi-channel radiances of different features tend to cluster in different places in the corresponding multi-dimensional space. This allows the computer to calculate the classes present and the resolution elements which belong to them. At a later stage the classes can be identified by reference to new ground truth, or through comparisons of sample data with a bank of spectral signatures. The particular advantages of such a method are that it is quick and cheap, no prior knowledge of the site is required, human bias is minimized, and future ground control stations can be located more objectively than in the 'supervised' techniques.

8.4 Operational, automatic decision/classification techniques

Often in the early stages of any remote sensing programme some retracing of one's steps is unavoidable in the march towards an operational feature recognition system. In the early, research, stages most attention is accorded to picture analysis rather than feature recognition. Often the best methods of data processing and analysis can be established only by trial and error, by the comparison of results obtained by different means. The expenditure of time, money and effort, and the complexity of the computer hardware and software to complete each method of approach must be assessed in terms of the resources which would be available operationally, and in terms of the requirements of the consumer judged by the additional criteria of accuracy, frequency and resolution.

Once the appropriate method of data analysis has been established the chief aim of most operational programmes is the correct recognition, identification, or classification of the contents of fresh imagery (Plate 8.1, see colour plates). Numerical techniques are superior to manual or photographic techniques in that the decision and classification rules are, by definition, quantitative or statistical, whereas those for the human interpreter are judgmental and substantially qualitative. Against this there is the disadvantage of automatic techniques that allowances are difficult to make for phenomena or conditions which are unevenly distributed or are present only from time to time. Such 'environmental noise' can be edited out by manual interpretation techniques, but may yield spurious patterns or results in automatic procedures.

Two cases in point are background brightness, which appears as an unwanted feature in cloud imagery from weather satellites, or foreground brightness related to clouds in Earth surface investigations. In satellite meteorology cloud cover maps (nephanalyses, see Chapter 9) are prepared routinely by hand to show, amongst other things, the percentage cloud cover categorised by broad classes. Today, automatic techniques for

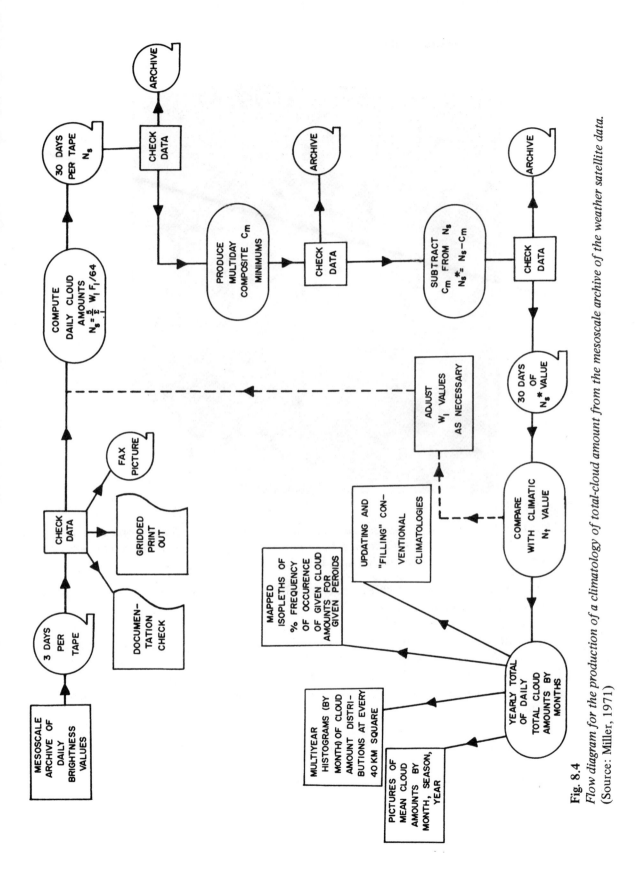

Fig. 8.4
Flow diagram for the production of a climatology of total-cloud amount from the mesoscale archive of the weather satellite data.
(Source: Miller, 1971)

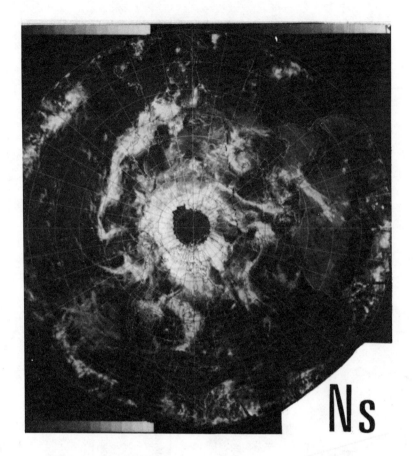

Plate 8.2

(a) The total Earth-cloud scene represented in the mesoscale resolution of computer-rectified, mapped imagery from operational weather satellites. Note the high brightness of the Arctic ice and snow and the Saharan sands as well as the clouds. (April 15th, 1969).

mean cloud cover mapping (see Fig. 8.4) include a stage at which the minimum brightness level for each picture point through the period in question is considered to be due to the albedo of the surface of the Earth. This is especially high in frozen regions, over sandy deserts, and where sun glint from water surfaces is strong. These brightness minima are subtracted from the daily mean brightness levels for every picture point to leave a value more representative of mean cloudiness (Plates 8.2a and b). Unfortunately it is more problematical to apply a technique like this to mapping of cloud cover on a short-term (e.g. daily) basis, and the manual method still prevails for everyday use.

At the end of any decision-making, classificatory procedure, an output is produced. This may be in one of a variety of forms, including graphical, cartographic, tabular modes.

(a) Graphical outputs. These usually consist of histograms of the frequency of occur-

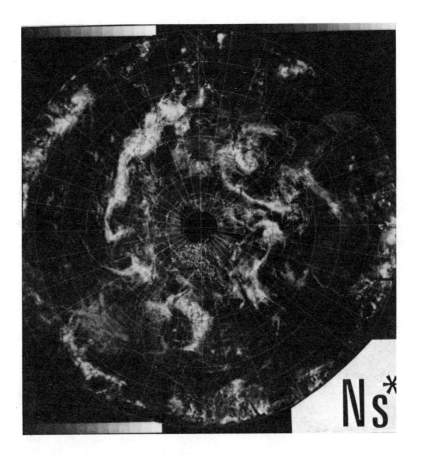

(b) The cloud scene for April 15th, 1969, remaining when the 30-day composite minimum cloud amounts were subtracted from the total Earth-cloud scene. (Source: Miller, 1971) (Courtesy, U.S. Air Weather Service, Scott Air Force Base, Illinois.)

rence of identified categories of features, or related categories.

(b) Map outputs. Maps are common, popular, long-hand or computer products in environmental remote sensing programmes. They may be qualitative (showing types of features, e.g. crops), quantitative (showing values of features or conditions, e.g. surface brightness: see Plate 8.3) or selected combinations of the two. The mapped patterns may be isoplethed or choroplethed, they may be single value or multiple category presentations, they may be based on absolute values, ratios, percentages, and changes through time; they may depict only point, line or area features, or all three simultaneously. In short, maps are highly versatile forms of digital programme outputs, and are justly significant in remote sensing as a whole.

(c) Tabular outputs. Once again a variety of forms and contents are possible. For example, the accuracy of identification may be summarized in matrix form, giving

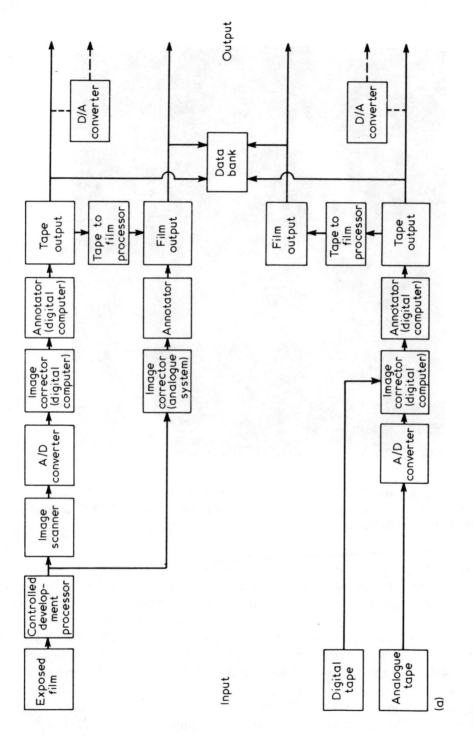

Fig. 8.5
Aspects of an automatic feature recognition facility for centrally processing data from ERTS satellites. (a) A data pre-processing facility.
(Source: EMI, 1973)

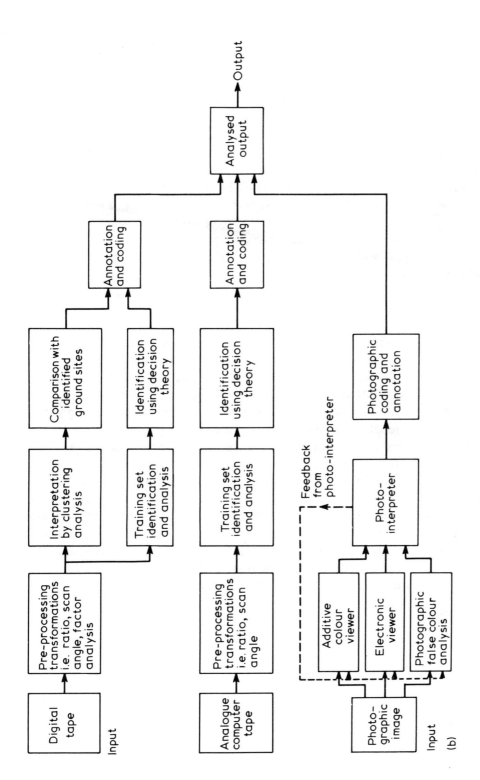

Fig. 8.5
Aspects of an automatic feature recognition facility for centrally processing data from Landsat satellites. (b) A feature recognition facility.
(Source: EMI, 1973)

correct identifications on the diagonals, and errors of omission and commission on the x- and y-axes. Such tabular outputs can be prepared easily by computer. This may form part of the checking ('verification') procedure by which the programme results are assessed against those obtained by other, especially conventional, methods. Any remote sensing programme for the analysis and interpretation of the target area stands or falls finally on its ability to provide worthwhile, especially cost-effective, results. Verification techniques often lead to the improvement of the procedures of analysis and interpretation by highlighting the errors.

We may conclude this necessarily brief and superficial introduction to the practices and problems of numerical processing by reference to an automatic feature recognition facility of the type proposed for central processing of ERTS satellite data. Figs. 8.5 a and b illustrate the data flow patterns through each part of the total facility, not only for digital and analog inputs, but also for inputs in a photographic form. After the initial simple image corrections have been made, and geometric and radiometric images have been reduced to a minimum, the data may be annotated, and a central store (the data bank) compiled (see Fig. 8.5 a).

In the feature recognition facility (Fig. 8.5b) the flow pattern is basically the same. Variations occur in the digital format due to the different forms of recognition techniques in use — basically the supervised and the unsupervised approaches. For photographic image inputs three forms of interpretation methods are shown, all of which require a photo-interpreter to analyse the results, and, if necessary change the processing parameters. This consequent feedback is represented by the dotted line.

Although much time and money has been expended in the development of feature recognition facilities, no truly automatic feature recognition complex has yet been constructed. Many recognition problems are very involved, on account of the variety and variability of components of the natural environment, and the factors which have a bearing upon them. The problem is exacerbated by the fact that many natural features — unlike many man-made features — have diffuse outlines, and irregular shapes. These features are especially difficult to identify and classify by objective methods, and human, subjective, and relatively qualitative methods will play an important part alongside quantitative techniques in remote sensing for some time to come.

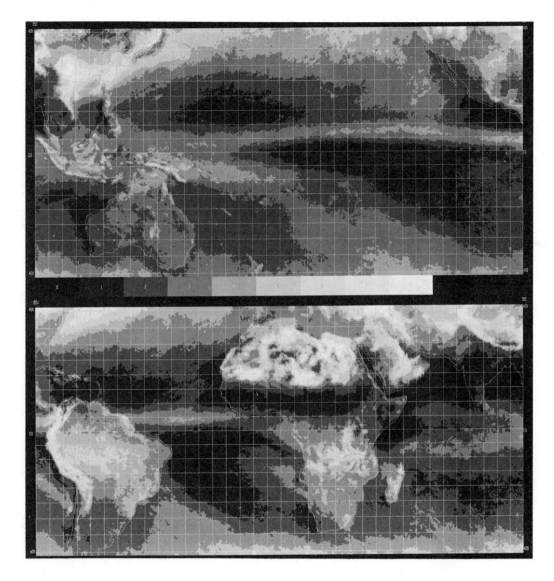

Plate 8.3
Photographic computer maps from ESSA satellites, averaged by seasons for a four year period. The use of the Mercator projection facilitates studies of tropical brightness patterns in general and trans-equatorial links in particular. Background brightness has been retained, hence, for example, the brightly-reflective desert areas of the Old World. (Courtesy, NOAA, 1972, (a): Dec, Jan, Feb., 1967-70).

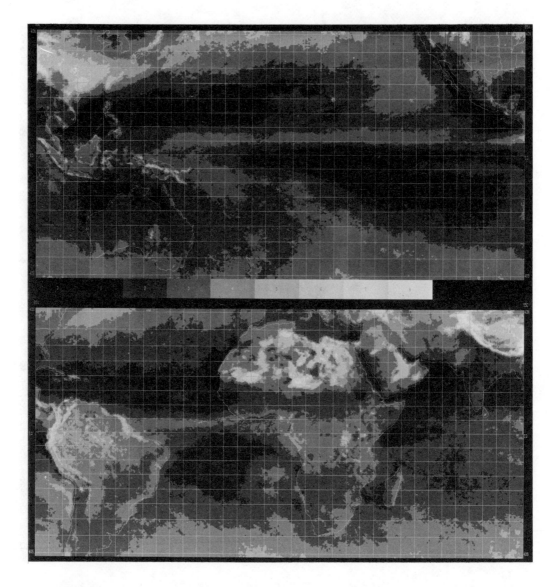

Plate 8.3
(b) March, April and May 1967–70. Less bright cloud appears in the tropics in this season than in the other three; the climatological reasons for this are not known.

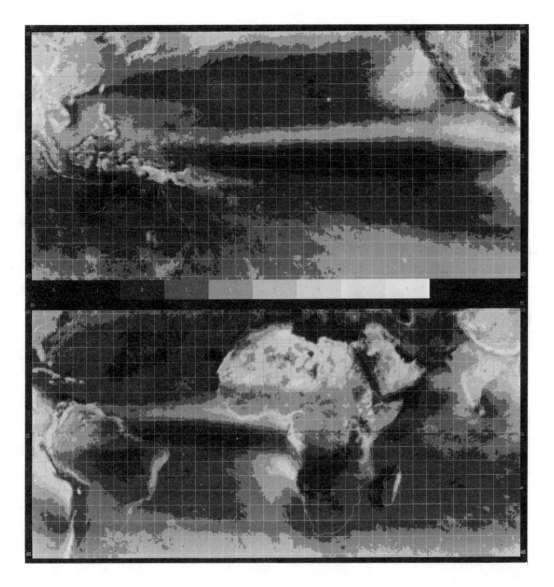

Plate 8.3
*(c) June, July and August 1967–70. The south Asian summer monsoon cloudiness
is particularly striking.*

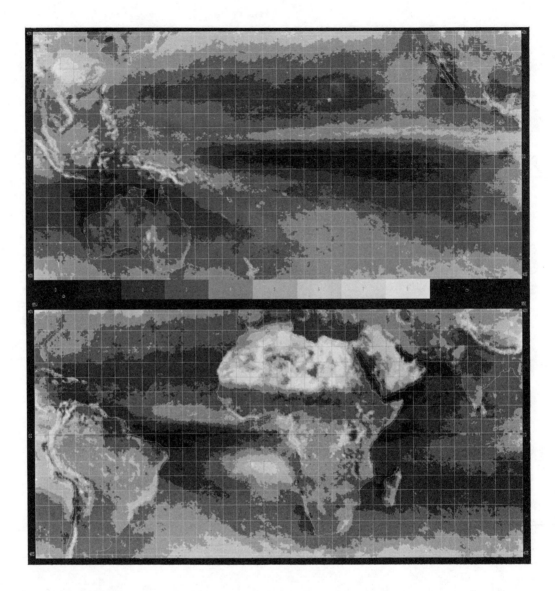

Plate 8.3
(d) September, October and November 1967–70. The constrasts between these patterns and those portrayed by Plate 8.3(b) are surprisingly large.

References

Bristor, C.L. (1968), 'Computer processing of satellite cloud pictures', ESSA Technical Memorandum, NESCTM-3, U.S. Dept. of Commerce, Washington.

Conlan, E.F. (1973), 'Operational products from Itos scanning radiometer data', NOAA Technical Memorandum, NESS 52, Washington.

EMI (1973), *Handbook of Remote Sensing Techniques,* D.T.I. Contract No. K46A/59, Technical Reports Centre, Orpington, Kent.

Miller, D.B. (1971), 'Automated production of global cloud climatology based on satellite data', Air Weather Services (MAC), Technical Report 242, Air Force, pp. 291-306.

Sabatini, R.R., Rabchevsky, G.A., and Sissala, J.E. (1971), 'Nimbus Earth resources observations', Technical Report No.2, Contract No. NAS 5-21617, Allied Research Associates Ltd., Concord, Massachusetts.

Simonett, D. (1974), 'Quantitative data extraction and analysis of remote sensing images', in *Remote Sensing, techniques for environmental analysis,* Estes, J.E., and Senger, L.W., (eds.), Hamilton Publishing Company, Santa Barbara, pp. 51-82.

Taylor, J.I. and Stingelin, R.W. (1969), 'Infrared imaging for water resources studies', *J. Hydraulics Division, Proceedings of the American Society of Civil Engineers,* **95**, 175.

Tryon, R.C., and Bailey, D.E. (1970), *Cluster Analysis,* McGraw Hill, New York.

Viglione, S.S. (1970), 'Applications of pattern recognition technology', in *Adaptive, Learning, and Pattern Recognition Systems,* Mendel, J.M., and Fu, K.S. (eds.), Academic Press, New York, pp. 115-162.

Weather analysis and forecasting

9.1 Remote sensing of the atmosphere

9.1.1 Advantages of remote sensing systems for weather studies

It is legitimate to ask why remote sensing should be welcome in support, or even in place, of conventional weather observation techniques which are well-developed, and long-established in many places. This question is vital, especially since the cost of a particular remote sensing instrument can be high in comparison with an *in situ* sensor. The advantages of remote sensing systems for weather studies include the following, although, of course, not all may apply necessarily to a particular investigation:

(a) The sensor does not need to be carried into the medium which is to be measured.

(b) The measurement system does not modify the parameter being measured, since this, is by definition, remote from it.

(c) Usually a high level of automation can be achieved with minimal effort.

(d) It is often possible to scan the atmosphere by remote sensing means in two to three dimensions unlike the single point measurement capability of most *in situ* systems.

(e) Integration of a given parameter along a line, over an area, or through a volume, is often obtained readily in the direct output of a remote sensor.

(f) Sophisticated and elusive parameters, such as the spectrum of turbulence and momentum flux, may be available as direct outputs of certain systems.

(g) High resolutions in time and space can be achieved with many types of remote sensing systems.

The three most common platforms for atmospheric remote sensors are the ground, aircraft (Plate 9.1) and satellites. The last have demonstrated great potential as vehicles both for operational meteorology and research. Much of the discussion that follows deals with the involvement of satellite systems in modern atmospheric science. Ground-based systems are generally less well-developed, although they have a bright future in meso- and microscale investigations. Our account will dwell mostly on radar as an example of a ground-based system, in view of its relatively broad and varied development to operational or near operational levels. Aircraft-based systems are the least cost-effective. Their almost entirely research-orientated applications have no significant place in an introductory review such as this book.

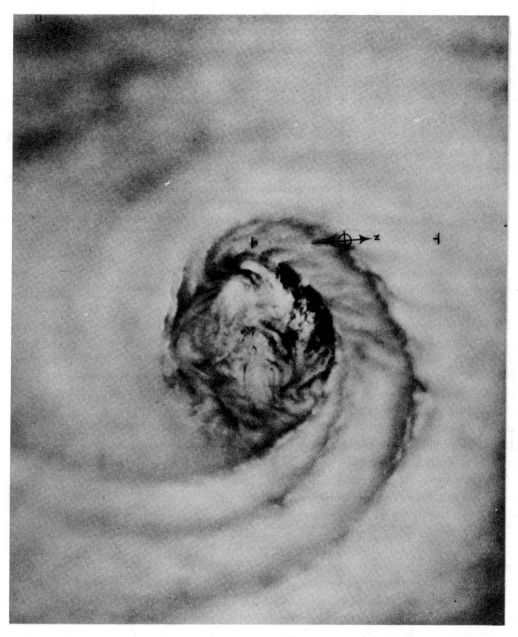

Plate 9.1
Typhoon Ida, September 24th, 1968, photographed from the lower stratosphere by a U-1 reconnaissance aircraft. The wall clouds around the eye have a marked helical arrangement. (Courtesy USWB.)

9.1.2 Ground-based remote sensing systems

Remote sensing of the atmosphere from the gound has been, and still is, a fragmentary affair. We shall see later that by far the greatest volume of atmospheric remote sensing data obtained today is gathered from a relatively small number of weather satellites. Two factors have restricted the development of ground-based techniques:

(a) Most of the atmosphere is invisible within the light waveband of the electromagnetic spectrum. Consequently (acoustic sounding devices apart), the state and structure of the atmosphere could not be assessed by instrumental remote sensing methods until passive infrared and active microwave systems had been sufficiently developed.

(b) The atmosphere is, from man's usual point of view, an enveloping, ubiquitous medium which, for many purposes, can be observed and measured satisfactorily from within. Conventional weather observatories are immersed in the atmosphere, and, though subject to certain limitations, provide data of the types required by standard forecasting procedures. However, this compatability has been, to a large extent, a child of necessity. Data from satellites are now adding a new dimension to well-established practices such as short-term weather forecasting even in areas generously supplied with conventional stations and are completing the scene where such stations are sparse.

The earliest form of artificially assisted remote sensing of the atmosphere was cloud photography from the ground. Later, balloons and aircraft were employed to provide new views of these very significant aggregates of water droplets. The invention of films of different speeds and sensitivies, filters to eliminate specific wavelengths of radiation, and time-lapse techniques in photography have all made important contributions to the meteorology of clouds. However, there are limits to the value of the essentially local information such studies of the clouds can provide. Cloud photography today is dominated by the amateur enthusiast, not the professional weather observer. In synoptic weather forecasting three aspects of the clouds are routinely observed:

(a) The proportion of the sky which is cloud covered.

(b) The type or types of clouds present at different levels.

(c) The height(s) of the cloud base (or bases if the cloud is multilayered).

Of the three the last is the only one which is evaluated sometimes by instrumental remote sensing from the ground. The height of the cloud base may be assessed either by searchlights (ceilingometers) or lasers.

Without doubt the most important ground-based system of atmospheric remote sensing — and that in most widespread use — is radar. This active microwave system was quickly developed for such purposes after the end of World War II. It may be used to identify and examine a wide range of atmospheric phenomena, including raindrops, cloud droplets, ice particles, snowflakes, atmospheric nuclei, and regions of large index-of-refraction gradients. Basically, the system consists of a transmitter, which produces power at the selected frequency, an antenna which radiates the energy and intercepts reflected signals, a receiver, which amplifies and transforms the received signals into video form, and a screen on which the returned signals can be displayed.

Often it is sufficient to know the strength of the echo from a target, and its distribution. No account need be taken of the frequency of the returning radar wave by comparison with the transmitted wave. Hence 'non-coherent' radar systems are adequate. For other studies, it may be important to know the direction and rate of movement of a phenomenon with respect to the radar location. In such cases the phase of the received signal must be compared with the transmitted signal. In other words, a 'coherent' radar system is required. One class of radar sets referred to as coherent (or Doppler) radars are designed to use the 'Doppler shift'. This is the change in frequency with which energy reaches a receiver when the receiver and the source of reflection are in motion relative to each other. Doppler radar is used in meteorology for a wide range of purposes, including the measurement of turbulence, updraft velocities, and atmospheric particle size distributions through tracking natural scatterers, and wind speed, through tracking artificial targets like rawinsonde balloons.

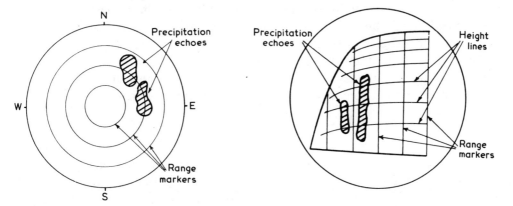

Fig. 9.1 (a) and (b)
The more commonly-used forms of radar displays, (a) the plan position indicator (PP1)*; (b) the range-height indicator* (RH1), (Source: Battan, 1973)

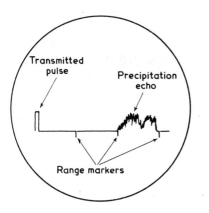

Fig. 9.1 (c)
The more commonly-used forms of radar displays, (c) the A-scope. (Source: Battan, 1973)

Some of the principal applications of non-coherent radars may be listed and elucidated in conjunction with various radar display units. These include:

(a) The identification and distribution of rain areas, and the recognition of the types
of rainfall included in them, on Plan Position Indicator (PPI) display units (see
Fig. 9.1a). By photographing the radar screen at intervals the development and
movement of rain areas can be followed and predicted (Plate 9.2). PPI displays
are widely used in operational meteorology for short-term rainfall forecasting, and
in research, for example in time-integrated studies of radar echo patterns and
satellite cloud photographs and/or conventional data.

(b) The study of vertical profiles through the atmosphere. These may be obtained
by a vertically-scanning radar, whose results may be displayed conveniently on a
Range Height Indicator (RHI) (Fig. 9.1b). This, like the PPI, uses intensity
modulation, i.e., the intensity of the bright spots corresponds to the strengths of
the returned signals. The RHI system may be used to study such phenomena as
cloud growth and development, or, in cloud-free conditions, atmospheric wave
motions and turbulence, and even that most intriguing yet elusive phenomenon
known as 'clear air turbulence' (CAT).

(c) The examination of echo intensities in chosen directions, using a simple A-scope
(Fig. 9.1c). This presents the backscattered energy in profile form, and permits
the operator to compare the echo with the transmitted pulse, which also appears
upon the trace.

9.1.3 Aircraft-borne systems

For many years larger aircraft of both military and civilian varieties have been equipped
with weather radar for detailed in-flight corrections to their flight paths. Such corrections may be made, for example, to avoid unnecessary encounters with powerful
convective clouds of the cumulonimbus and cumulocongestus types. Here strong
thermal uplifts and pronounced atmospheric turbulence may occur.

In atmospheric research airborne radars of the noncoherent type with vertically
scanning antennae, or fixed, vertically pointed, antennae have been used to study
the development and spread of precipitation especially in severe weather systems like
hurricanes, and in association with active mid-latitude fronts. The idea of airborne
Doppler radars for meteorological puposes is relatively new, an added complication
accompanying the movement of the sensor platform as well as the target. We may
look forward to successful operations employing such systems before the end of the
present decade.

9.1.4 Satellite systems

Remote sensing of the atmosphere received a tremendous fillip with the launching of
the first specialized weather satellite in April, 1960. It was not until the satellite era
had opened that investigations of weather by remote sensing means became a practical

A summary of atmospheric satellite families, 1959-73

Family name	Country of origin	Number launched*	Approx. period covered	Special remarks
Vanguard	U.S.A.	1 **	Feb-Mar 1959	Early experimental satellites with
Explorer	U.S.A.	2 **	Aug 1959-Aug 1961	primitive visible and infrared imaging systems
Television and infrared observation satellite (Tiros)	U.S.A.	10	Apr 1960-July 1966	First purpose-built weather satellites
Cosmos	U.S.S.R.	22 **	Apr 1963-Dec 1970	Some weather satellites in this large, cosmopolitan Russian satellite series
Nimbus	U.S.A.	6	Aug 1964-	Principal American R & D weather satellite
Environmental survey satellite (Essa)	U.S.A.	9	Feb 1966-	First American operational weather satellite
Molniya	U.S.S.R.	8	Apr 1966-May 1971	Dual purpose communication/weather observation satellite
Applications technology satellite (ATS)	U.S.A.	4	Dec 1966-	First meteorological geostationary satellite to test SMS concepts
Dodge	U.S.A.	1	July 1967-Jan 1971	Tested gravity stabilization techniques, but also took first colour pictures of Earth and clouds
Meteor	U.S.S.R.	12	Mar 1969-	Current Russian operational weather satellite series
Improved Tiros observational satellite (ITOS)	U.S.A.	1	Jan 1970-June-1971	NOAA prototype
National oceanic and atmospheric administration satellite (NOAA)	U.S.A.	3	Dec 1970-	Current American operational weather satellite series
Péole	France	1	Dec 1970-	Eole prototype
Eole	France	1	Aug 1971-	
Defense meterological satellite program (DMSP)	U.S.A.	2	Feb 1973-	Military satellites.

* As of 30/6/73 (Meteorological satellites only: ** denotes families which have included other satellites designed for other basic purposes).

possibility on anything other than a very piecemeal, localized scale. Prior to 1960 a number of spacecraft and satellites had made some preliminary observations of the Earth's atmosphere from orbital altitudes, e.g. high-altitude rockets of the Viking series, and members of the American Vanguard and Explorer satellite families, but the first Tiros (Television and Infrared Observation Satellite) may be considered properly as the harbinger of satellite meteorology. By the time of writing about ninety satellites have been launched for the primary purpose of monitoring the Earth's atmosphere. Rather less than half of these have been 'operational' satellites intended to back up and extend the conventional weather observing facilities, and improve weather forecasts. Rather more than half have been 'research and development' (R & D) satellites, test-beds for new concepts, equipment and instruments. The principal atmospheric satellites concerned with observing weather patterns directly affecting man are listed in Table 9.1. Some other satellites have investigated solar phenomena and patterns in the upper atmosphere. In this account we need concern ourselves only with those which view man's immediate atmospheric environment — the weather in which he is immersed.

Since data from the American weather satellites are much more readily available than those from their Russian counterparts we shall concentrate the remainder of our discussion in this chapter and the next on the American programme and its results, except in certain special instances. The orbital configurations of the principal families are depicted in Fig. 9.2. All are relatively low altitude satellites orbiting the Earth several times each day at high angles to the equator, with the exception of the A.T.S. family and S.M.S., its recently inaugurated successor. These two types occupy geosynchronous orbits. Here the rate of precession of a satellite around each orbit is related to the spin of the Earth on its axis so that the satellites appear to hover above a fixed point on the Earth's surface.

The ideal global observing system during the late 1970's has been conceived along the lines of Fig. 9.3. Three or four geosynchronous satellites are required to give a complete coverage of low latitudes (effectively within 35–40⁰ north and south of the equator), imaging the Earth and its cloud cover at frequent intervals (every 24 or 12 minutes). Middle to high latitudes are difficult to view from the equator because of foreshortening caused by the curvature of the Earth. Consequently a pair of near-polar orbiting satellites is required to give at best a data coverage every 6 hours. This is generally met at present by two Noaa satellites. Fig. 9.3 shows that a low-altitude satellite of the Noaa family is included in the World Weather Watch scheme in an equatorial orbit to provide infrared data of a higher resolution than the geosynchronous satellites could provide easily from their lofty viewpoints, 35 400 km up. Provisions to complete such a system have been made by the World Meteorological Organisation through its operational programme codenamed World Weather Watch (WWW), in conjunction with its research programme, the Global Atmospheric Research Program (GARP).

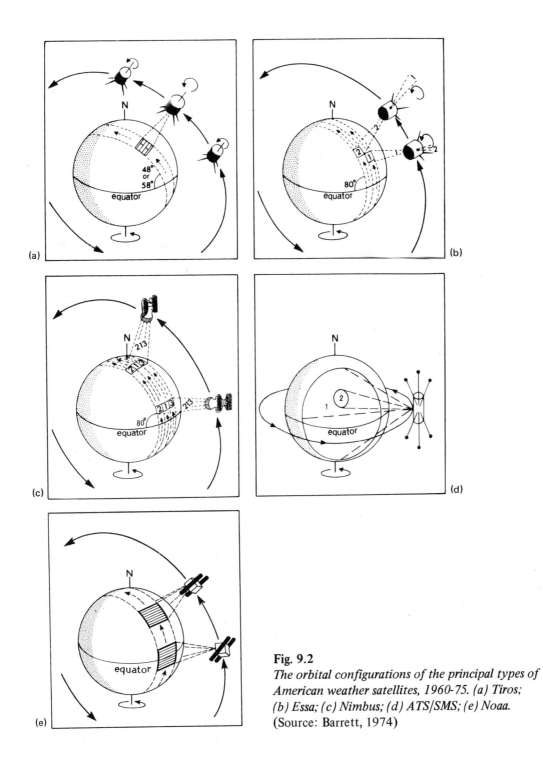

Fig. 9.2
The orbital configurations of the principal types of American weather satellites, 1960-75. (a) Tiros; (b) Essa; (c) Nimbus; (d) ATS/SMS; (e) Noaa.
(Source: Barrett, 1974)

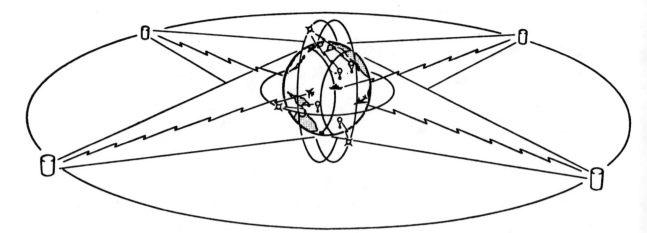

Fig. 9.3
A proposed global observing system for GARP, involving four geosynchronous satellites, three low orbiting satellites (one in an equatorial orbit and two in polar orbits), weather balloons, weather ships and instrumented aircraft in addition to conventional surface observatories. (Source: Suomi, 1970)

We may review the observational sensor systems flown on meteorological satellites past and present under three headings (Chapters 3 and 4) namely:
(a) Visible imaging systems.
(b) Infrared radiometers.
(c) Sounding spectrometers.
It will be useful to discuss these briefly in turn before we proceed to consider the analysis and interpretation of the data they provide for meteorological uses. Climatological uses will be described in Chapter 10 and hydrological programmes in Chapter 11.

(a) Visible imaging systems. Of these, the great majority have been of four broad types, namely Vidicon Camera Systems (VCS), Image Dissector Camera Systems (IDCS), Spin-Scan Cloud-Cover Camera systems (SSCC), and visible waveband radiometers. In general, the resolutions of these systems at nadir have been of the order of a few kilometres. This is deemed satisfactory for synoptic forecasting purposes; though research workers would doubtless like at least some data of a higher resolution. This need is now being met by the visible channel (0.6–0.7 μm) of the Very High Resolution Radiometer (VHRR) on Noaa, giving a nominal resolution of 0.9 km. The vidicon camera systems were employed on Tiros satellites, and were replaced on Nimbus and Essa (the first operational satellites) by advanced (AVCS) developments. The principal differences were in resolution and the number of cameras comprising the system. VCS cameras operated individually, AVCS cameras in pairs, or arrays of three. All took snapshot photographs, their shutters opening to expose a light-sensitive vidicon tube to reflected light from the target area. Between exposures the pictures were read off the vidicon tubes and were either stored before play-back to a Command and Data Acquisition (CDA) station in the U.S.A., or, in some cases, were transmitted immediately for any simpler Automatic Picture Transmission (APT) station in radio range to intercept. On the ground the pictures were displayed on television screens at CDA stations and photographed by 35 mm cameras to provide a permanent record, or were reconstructed on facsimile machines at APT stations.

The IDCS systems, employed on some Nimbus and ATS satellites, consisted of shutterless electronic scans which built up pictures line by line on a photocathode tube before transmission of the data to Earth.

The SSCC systems included a high-resolution telescope and (in the Multicolor, MSSCC, version) three photomultiplier light detectors (red, blue and green), plus a precision latitude step mechanism to provide pole-to-pole coverage in 2400 scan lines as the geosynchronous ATS platform spun on its axis. A total time of 24 minutes was required to scan one frame with a nominal satellite rotation of 100 revs/min. The most significant feature of this system by comparison with those outlined earlier is its ability to provide colour images, although the extra utility of colour for purposes of cloud analyses is rather doubtful (Plate 9.3 see colour plates).

Several types of visible waveband radiometers have been flown by different weather satellites. The first was the visible channel sensor in the Medium Resolution Infrared Radiometer (MRIR) flown by certain Tiros and Nimbus satellites. The MRIR provided data with a resolution generally no better than about 40 km. These were helful in orientating the infrared waveband data correctly. More importantly, current Noaa satellites incorporate a Scanning Radiometer (SR) system consisting of two sensors and their ancillary equipment. One operates in the $10.5-12.5$ μm atmospheric window waveband for day and night viewing, the other in the visible waveband from $0.52-0.73$ μm for day time imaging alone. (See Plate 9.4). The scanning radiometer forms an image using a continuously rotating mirror which scans the Earth's surface perpendicular to the satellite's orbital path. As the satellite progresses along its orbital path each rotation of the mirror provides one scan line of picture. Radiation collected by the mirror is passed through a beam splitter and spectral filter to produce the desired spectral separation. The data can be stored, or transmitted in real time for reconstruction on a facsimile recorder (Fig. 9.4). The VHRR, mentioned above, is similar, but improves significantly on the 4 km resolution of SR at nadir.

(b) Infrared radiometers. For the sake of convenience we may consider these to be different from spectrometers in that they are intended to provide data relating basically to the horizontal plane, whereas the purpose of spectrometers is to provide data which can be used directly to reveal patterns in the vertical plane. In other words, we may say that infrared radiometers are intended primarily to examine the atmosphere in breadth,

Plate 9.4 (a)
*Examples of computer-rectified, brightness-normalized, imagery from a Noaa operational satellite.
(a) and (b) are visible images for the northern and southern hemispheres, (c)-(f) infrared images for
the same day. Infrared imagery can be obtained by night as well as by day.* (Courtesy, NOAA)

whereas spectrometers examine it in depth. We have already discussed and illustrated
the operation of one such instrument in some detail, namely the THIR. (See pp. 36–37).
The MRIR-type of radiometer was summarized on p. 35. Good quality analog signals
and images have been retrieved from the High Resolution Infrared Radiometers (HRIR)
of Nimbus satellites (about 3 km resolution at nadir), and, more recently, from the new
Very High Resolution Radiometers (VHRR) of Nimbus and Noaa satellites (circa 0.9 km
resolution). (See Plate 9.5).

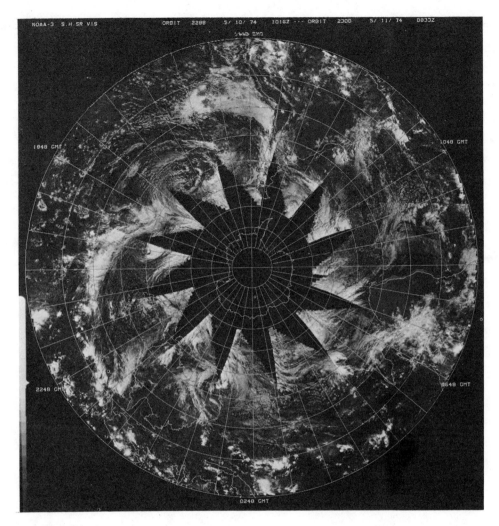

Plate 9.4 (b)

Radiation measurements made simultaneously in a number of carefully selected wavebands have provided much information concerning the state and behaviour of the atmosphere, especially pertaining to the radiation balance of the Earth/atmosphere system.

(c) Sounding spectrometers. This group, with its particularly useful application in profiling the atmosphere in depth, is rapidly expanding. Some, like the Satellite Infrared Spectrometer (SIRS), determines the vertical temperature and water vapour profiles of the atmosphere through a fixed grating. Infrared radiation received by the sensor is measured in 13 selected spectral intervals in the CO_2 and water vapour absorption bands,

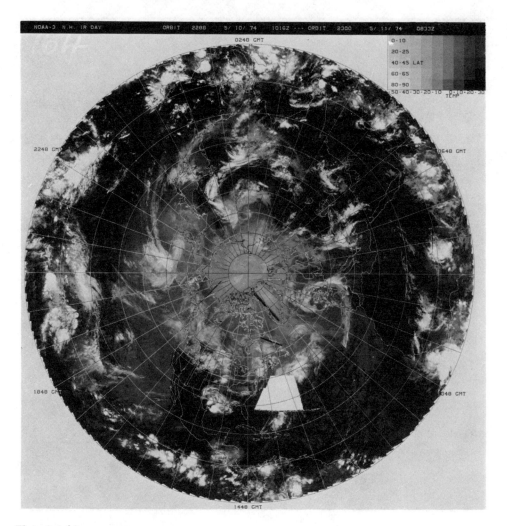

Plate 9.4 (c)

and in one channel in the 11 μm atmospheric window. Successful temperature profiles have been achieved using SIRS data, and these have been employed to correct upper tropospheric synoptic charts for forecasting purposes (see p. 172). Another type is the Infrared Interferometer Spectrometer (IRIS). This measures Earth surface and atmospheric radiation in the broad waveband from 6.25–2.50 μm using a modified Michelson interferometer. Radiation from the target is split into two approximately equal beams to give an interferogram which is eventually transmitted to the ground. Here a Fourier transform is performed to produce a thermal emission spectrum of the Earth (Fig. 9.5). From the water vapour, ozone and carbon dioxide absorption bands in the spectrum vertical profiles of the concentrations of these gases in the

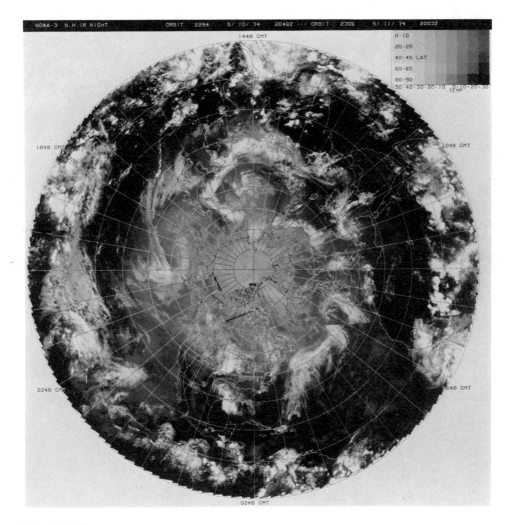

Plate 9.4 (d)

atmosphere can be derived. Several other types of weather satellite spectrometers are planned for the near future, and this seems likely to become one of the most important growth areas in the science of weather study from space.

9.2 Satellite data applications and operational meteorology

9.2.1 General forecasting applications

The principal use of visible imagery from satellites has been through the analysis of the cloud content to provide an additional input into the pool of data for short-term

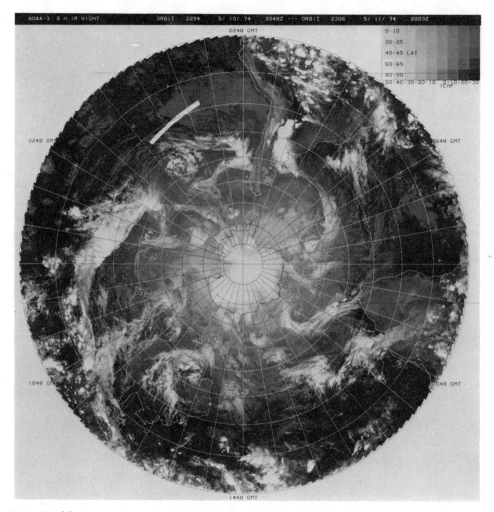

Plate 9.4 (e)

weather forecasting. Over much of the surface of the Earth conventional meteorological stations are still too thinly spread to meet the requirements of modern forecasting programmes. Weather satellite data are especially useful in the conventional data-remote regions, which include much of the tropics, most ocean areas in middle latitudes, and the polar regions.

It was established soon after the first weather satellite had been launched that many of the species of clouds in standard classifications could be identified in satellite imagery under reasonable conditions. Table 9.2 summarizes the characteristics of clouds photographed from satellites whereby the principal cloud types may be identified.

For several years it has been standard practice in the forecasting offices of many nations around the world to prepare simplified cloud charts (satellite nephanalyses)

Plate 9.4 (f)

from the available visible imagery — generally A.P.T. pictures in countries other than
the U.S.A. Fig. 9.6a summarizes the standard nephanalysis key, whilst Fig. 9.6b is an
example of one such chart. The information content includes cloud type, the strength
of cloud cover, an indication of the structure of the cloud fields, interpreted features
such as cyclonic vortex centres and jet streams, and boundary lines for ice and snow
as well as clouds. It has been estimated that the information content of a nephanalysis
is two orders of magnitude less than that of the image on which it is based. This is
usually sufficient for the purposes of the short-term forecasters, who may compare the
nephanalyses with the conventionally-based synoptic charts for the day, and correct
these where necessary in the light of the satellite information. Especially where the
forecasting area includes substantial sea areas and coastal regions the satellite evidence

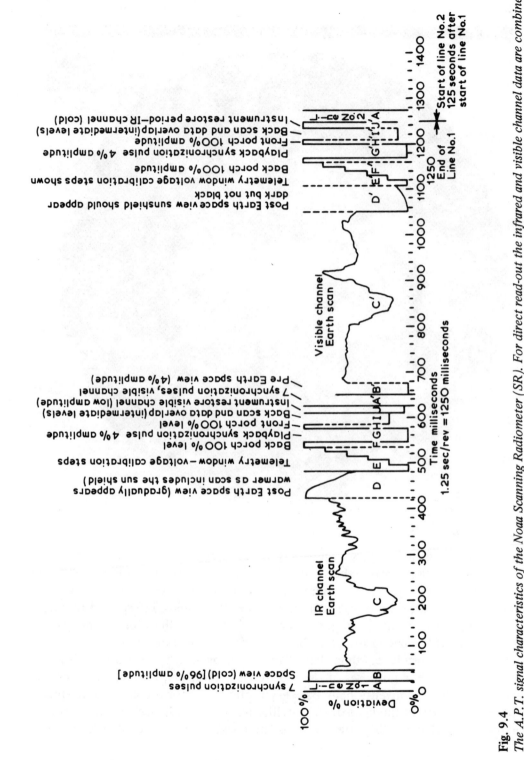

Fig. 9.4
The A.P.T. signal characteristics of the Noaa Scanning Radiometer (SR). For direct read-out the infrared and visible channel data are combined in a single data stream. (Source: Schwalb, 1972)

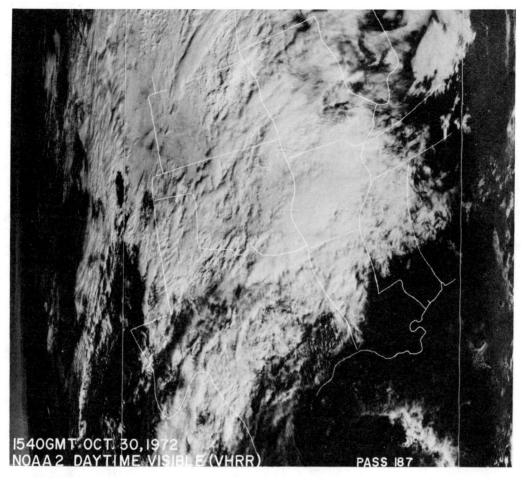

Plate 9.5

A Very High Resolution Radiometer (VHRR) view of part of the U.S.A. (Courtesy, NOAA)

may be the salvation of a forecast through the correct identification of, say, approaching fronts, or ridges of clear weather. Current computer-based forecasting methods provide maps of the anticipated contour patterns for selected pressure surfaces; frontal analyses are still carried out by hand. It is in this respect, and the related respects of cloud cover and rainfall forecasting, that weather satellite visible imagery is most valuable.

Since the operational Essa satellites began to be replaced by the Noaa family at the end of 1970 it has become possible for APT stations to receive data for their local regions in the infrared as well as the visible waveband. This has opened up a new range of possibilities, for example, the opportunity to compile twice daily nephanalyses. However, perhaps the most important single application of infrared data in weather

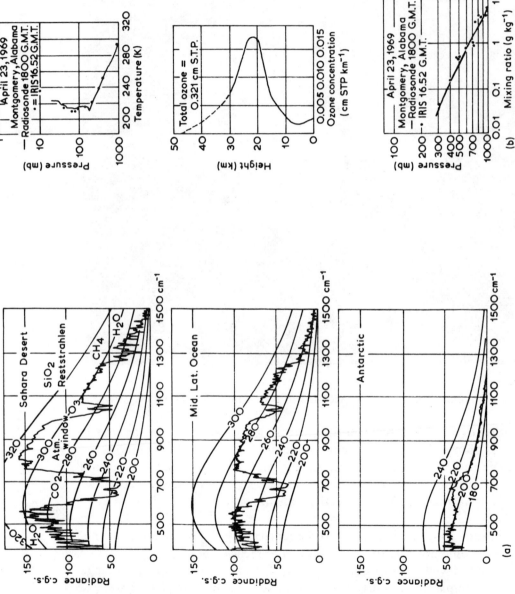

Fig. 9.5

Data received from the Nimbus Infrared Interferometer Spectrometer (IRIS) experiment have been processed to give: (a) Thermal emission spectra of the Earth from 5.25 μm. Here representative spectra on one day for three regions of the world are compared with blackbody curves. The departures are due to both atmospheric absorption, and changes of emissivity of the source with wavelength. At these wave numbers the primary absorption constituents of the atmosphere are water vapour (H₂O), carbon dioxide (CO₂), ozone (O₃), and methane (CH₄). In the broad atmospheric window surface temperatures in the three graphs (from top to bottom) are 312K, 285K and 200K respectively. (b) Temperature, humidity and ozone profiles. Excellent correspondence is found between these profiles and radiosonde data and measurements from ground-based spectrometers. In sharp contradistinction to data from the scatter of surface weather stations IRIS returns continuous strips of

Table 9.2
Characteristics of clouds portrayed by satellite visible images

Cloud type	Size	Shape (organization)	Shadow	Tone (brightness)	Texture
Cirriform	Large sheets, or bands, hundreds of km long, tens of km wide.	Banded, streaky or amorphous with indistinct edges.	May cast linear shadows esp. on underlying cloud.	Light grey to white, sometimes translucent.	Uniform or fibrous
Stratiform	Variable, from small to very large (thousands of square kms).	Variable, may be vertical, banded, amorphous, or conforms to topography.	Rarely discernible except along fronts.	White or grey depending on sun angle and cloud thickness.	Uniform or very uniform.
Strato-cumuliform	Bands up to thousands of km long; bands or sheets with cells 3–15 km across.	Streets, bands, or patches with with well-defined margins.	May show striations along the wind.	Often grey over land, white over oceans, due to contrast in reflectivity.	Often irregular, with open or cellular variations.
Cumuliform	From lower limit of photo-resolution to cloud groups, 5–15 km across.	Linear streets regular cells, or chaotic appearance.	Towering clouds may cast, shadows down sunside.	Variable from broken dark grey to white depending mainly on degrees of development.	Non-uniform alternating patterns of white, grey and dark grey.
Cumulo-nimbus	Individual clouds tens of km across. Patches up to hundreds of km in diameter through merging of anvils.	Nearly circular and well-defined, or distorted, with one clear edge and one diffuse.	Usually present where clouds are well-developed.	Characteristically very white.	Uniform, though cirrus anvil extensions are often quite diffuse beyond main cells.

CLOUD TYPES

Cumuliform   Strato-cumuliform

Apparent Cumulo-nimbus   Stratiform   Cirriform
or Cumulo-congestus

CLOUD AMOUNTS (% COVER)

O	—	Open	—	< 20%
MOP	—	Mostly open	—	20–50%
MCO	—	Mostly covered	—	50–80%
C	—	Covered	—	> 80%

SIZES OF CLOUDS AND SPACES

Cloud	Size (nautical miles)	Open spaces
1	0–30	6
2	30–60	7
3	60–90	8
4	90–120	9

BOUNDARIES

)))))) Major cloud system   ———— Definite

+ + + + + Limit of ice or snow   – – – – Indefinite

PATTERNS AND SYNOPTIC INTERPRETATIONS

Vortex   Anticyclone centre

Comma-shaped   Wave clouds
cloud mass

Cloud line (form may be ⌢ , /// ,))) , ⋈)

Tenuous cloud line   Change of element size
along line as shown

⟵⟶ Striations   ⟵ — ⟶ Tenuous striations

⟹ Direction of   ➡ ➡ ➡ Estimated location
cirrus streakiness   of jet

+ Bright cloud mass   — Thin cloud mass
(transparent or translucent)

TERMS

Element	Cellular
Cloud mass	Eddy
Cloud pattern	Hazy
Cloud system	Probable
Cloud band	Possible

Fig. 9.6

(a) The standard nephanalysis key. (Source: Barrett, 1970)

forecasting arises from the possibility that 'three-dimensional nephanalyses' may be compiled therefrom (Plate 9.6). Data from snapshot radiometers or spectrometers like SIRS, and its successor in the operational Noaa satellites, the Vertical Temperature Profile Radiometer (VTPR) can be processed to give a range of useful products as illustrated by Fig. 9.7. Results have shown that satellite-derived vertical temperature profiles compare well with profiles obtained by radiosonde balloons. A regular

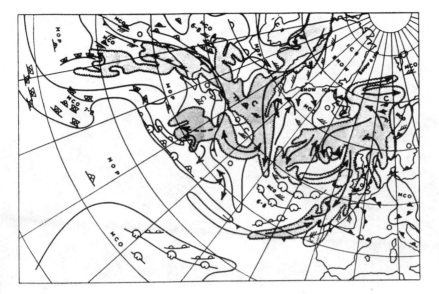

Fig. 9.6
(b) A sample standard nephanalysis summarizing cloudness over the North Atlantic, on July 17th, 1967. (Source: Barrett, 1970). An improved scheme has been suggested by Harris and Barrett, 1975.

pattern of sounding snapshots around the world can provide information of sufficient quality and quantity to support more accurate contour charts for selected surfaces than conventional means permit.

When we turn to review the operational applications of weather satellite data on a regional basis it is clear that weather forecasting in the tropics stands to derive particular benefits from such a source. Here the joint operation of low-altitude and geosynchronous satellites is specially advantageous. Although some indications of wind directions, and even speeds, can be derived sometimes from once or twice daily satellite data, frequent imagery of the type provided by ATS satellites is far superior in these respects. It has been shown that, by identifying particular clouds on successive photographs, ideally one to two hours apart, winds may be mapped on the basis of their displacements. Maps of so-called 'cloud motion vectors' can be analysed to give maps of streamlines and isotachs, which together form a complete analysis of the wind field (Fig. 9.8). Indeed, it has been established that both upper and lower wind fields may be mapped on the basis of such data, the one relating generally to the 300 mb level, and the other to 900 mb. Of course, once the maps of streamlines and isotachs have been compiled, a number of dependent maps can be drawn up. These include patterns of relative and absolute vorticity, and atmospheric divergence. Tropical meteorology will undoubtedly benefit enormously from the operational geosynchronous weather satellite system planned for the second half of the 1970's.

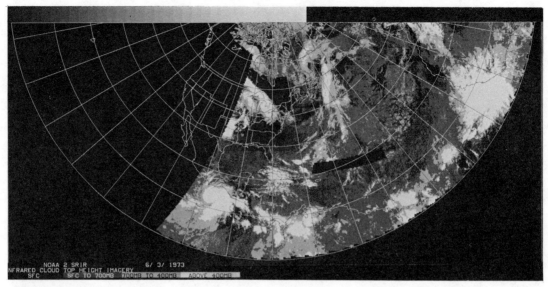

Plate 9.6
Noaa 2 infrared imagery generalized to indicate four temperature classes relating to surface temperatures, and cloud top temperatures within the height ranges surface − 700 mb, 700 mb−400 mb and above 400 mb. The poleward gradient of decreasing atmospheric thickness has been taken into account. (Courtesy, NOAA)

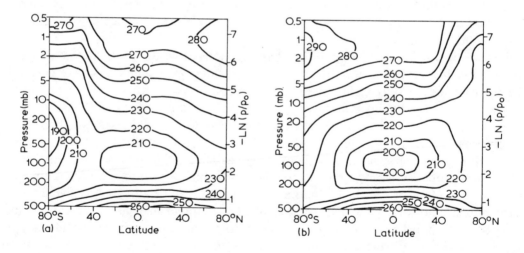

Fig. 9.7
*Meridional cross-sections of zonally averaged temperatures (Isopleths give temperature in K):
(a) 16th July 1970; (b) 21st January 1971, obtained with Selective Chopper Radiometer (SCR) on
Nimbus 4, and thickness of the 10−1 mb pressure surfaces from the same source: (c) Northern
hemisphere; (d) Southern hemisphere, both on 4 July, 1970,* (Source: Barnett *et al.*, 1973)

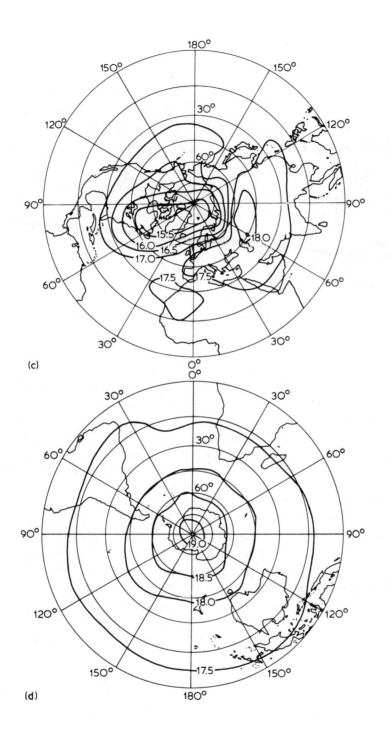

(c)

(d)

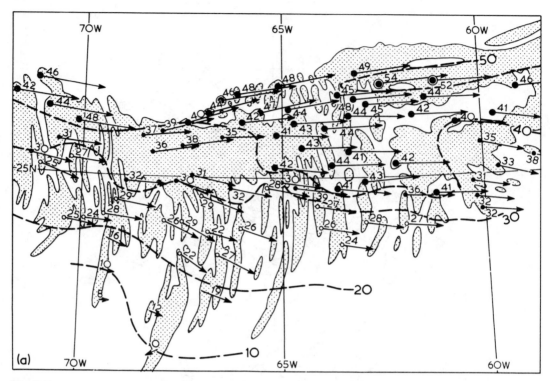

Fig. 9.8
(a) Cloud motion vectors, indicating the speed and direction of the upper tropospheric winds over part of the western North Atlantic (Bermudan region), drawn from a series of ATS satellite pictures on May 24th, 1968. The situation depicted (cloud areas stippled) was photographed at 1212Z. Cloud elements indicated by circles displaced at the indicated speeds (in knots) between 1212Z and 1405Z.

9.2.2 Special forecasting applications

In many forecasting situations certain weather structures are of particular significance, usually because extreme events are associated with them. Over the years, research in many centres, but most especially in the National Environmental Satellite Service of the U.S. Weather Bureau, has established that a very wide range of synoptic weather systems and structures can be identified in satellite imagery, either visible or infrared. Prominent amongst them are vorticity patterns. A wide range of tropical and extra-tropical cyclones of different types and at different stages of development have been recognized and modelled. Frontal forms, and also the results of frontogenesis and frontolysis, can often be identified. Various cellular cloud patterns are known to be related to different instability conditions. In the upper troposphere, upper level troughs and jet streams can be identified and classified. Anticyclones have always been more difficult to analyse since they are characteristically cloud-free, but the shapes and forms of relatively clear areas, and their fringing clouds, have been used to differentiate

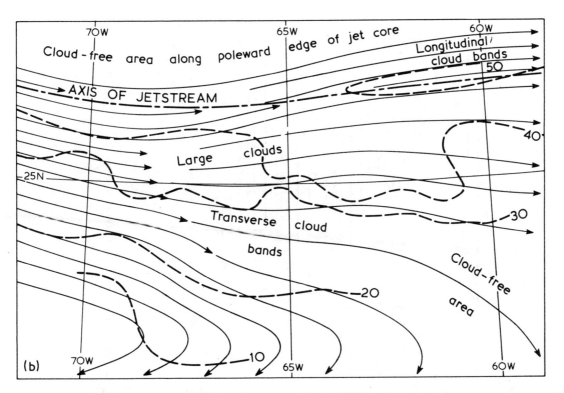

(b) Streamlines of cloud velocity (broken lines, quantified in KTS) and the isotach pattern, constructed from the information in Fig. 9.8 (a)

between different types of ridges. Lastly, many cloud patterns associated with particular combinations of atmospheric and topographic interactions have been recognized. Fig. 9.9a and b represents the basis of a cloud classification which involves not so much the types of clouds that can be identified as the relationships between cloud systems and the atmospheric structures with which they are associated. Many of the included forms suggest to the forecaster a particular likelihood or set of likelihoods so far as the development of the weather is concerned.

Certain of the weather systems embraced by Fig. 9.9 (a) are particularly destructive, and, as a consequence, have been studied with special care. Weather satellites have almost certainly justified their expense since the mid-1960's through the assistance they have given in hurricane forecasting alone. Probably no hurricane or tropical storm anywhere in the world has gone unnoticed for the last decade. During this time we have learnt much about the structures and movements of these small but powerful vortices from satellite evidence. Very detailed classifications have been compiled for satellite-viewed hurricane cloud patterns. One of these schemes has been used as the basis for estimating the maximum wind speeds associated with them (Fig. 9.10). More recently it has

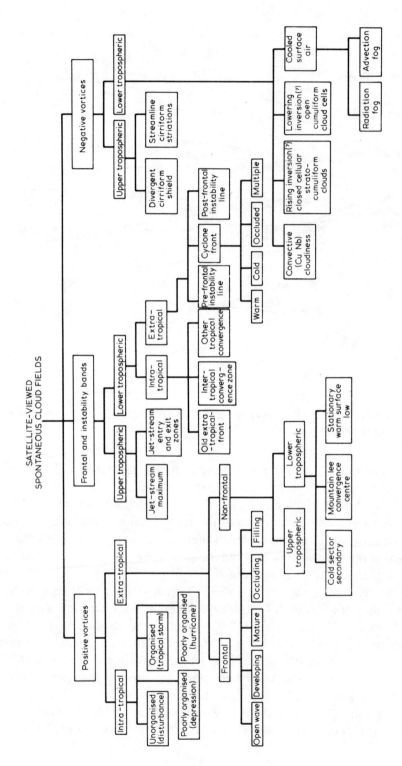

Fig. 9.9

(a) A basis for a genetic classification of cloud systems portrayed by satellite imagery. Section 1: systems organized dominantly by processes affected little by surface topography. (Source: Barrett 1970).

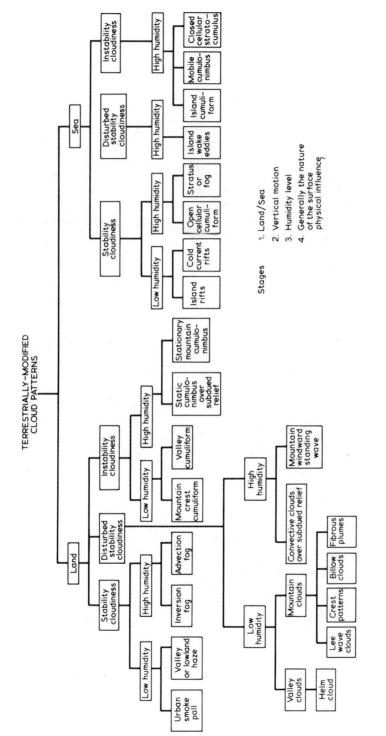

Fig. 9.9

(b) Section 2: cloud systems and cloud rift patterns prompted mostly by surface characteristics. (Source: Barrett 1970).

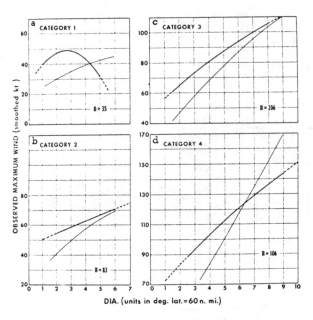

Fig. 9.10

Observed relationships between the size and shape of imaged hurricanes and their maximum wind speeds, including original (light line) and verified (heavy line) curves. Prolonged experience with predictive nomograms based on remote sensing data should be used in this way to verify, and, where necessary, improve original schemes. (Source: Hubert and Timchalk, 1969)

been suggested that wind estimates may be made from infrared data in digital or image form with equal ease. The damage caused by a single hurricane in the south-east corner of the U.S.A. has been known to exceed $ 1 500m. In November 1970 a tropical cyclone struck the head of the Bay of Bengal and the loss of life caused by the associated wind, rain, and tidal flooding exceeded 300 000 people. Indirectly, this was the trigger which led to the establishment of the independent nation of Bangladesh. Clearly any new information we can get about the behaviour of such significant storms may be almost priceless.

9.2.3 Research and development

There are two aspects of this matter, one involving new sensors and new modes of data analyses, the other searching for a new understanding of the structure and behaviour of the atmosphere. Continuing efforts are being made to improve existing satellite sensors, and to develop others which will provide more or better data. Most of the instruments flown by current operational satellites were tested first on Research and Development satellites like Nimbus. Others, for example, the Infrared Temperature Profile Radiometer (ITPR) of Nimbus 5, is being tested as a replacement for previous infrared radiometers and spectrometers, its higher spatial resolution allowing more accurate determinations of atmospheric temperature profiles in partly cloudy areas.

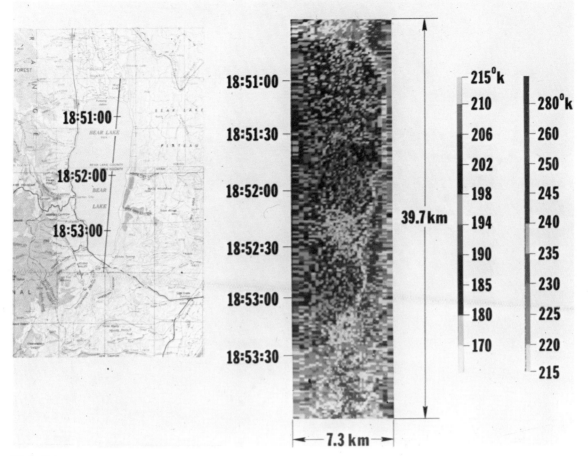

Plate 3.4

The flight path (3.4 (a)) and microwave image (3.4 (b)) (1.55 cm waveband) for March 3rd 1971, Bear Lake, Utah/Idaho, from an altitude of 3400 m. (See p45). (From Schmugge et al., 1973)

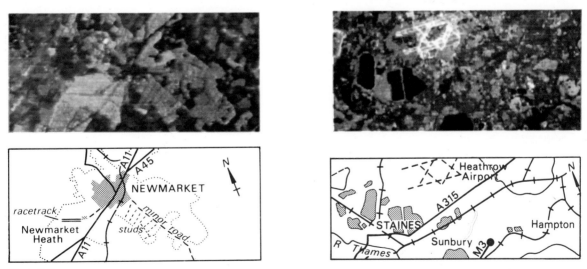

Plate 8.1

Examples of various land use categories obtained by computer processing of Landsat 1 data for the Department of Environment. The processing minimizes degradation of the picture detail. Scale of reproduction 1: 250 000. See key maps for selected features. (a) Urban area of Newmarket. Rural areas image in red and green. Newmarket Heath images in yellow. (b) Heathrow airport and adjoining areas. Rural areas are imaged in red and green, reservoirs in black and construction works in white.

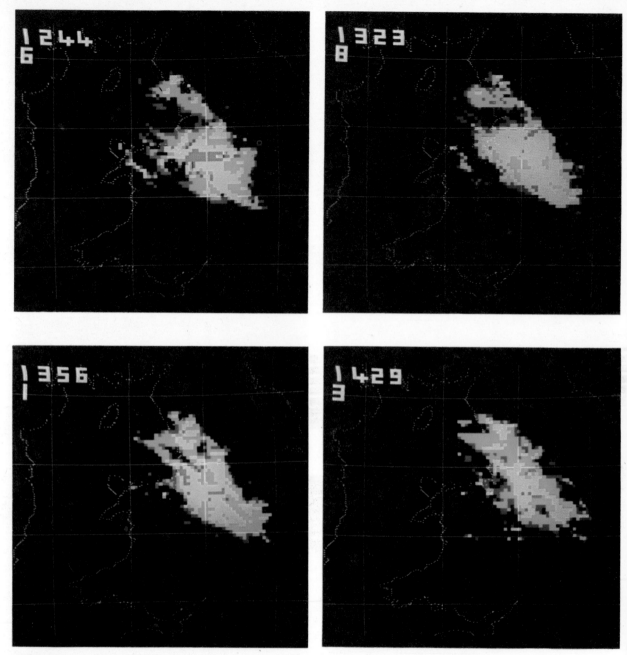

Plate 9.2

Time sequence of four photographs of a colour television display showing the distribution of rain ahead of a warm front on February 14th, 1974. Colours represent different rainfall intensities. Equipment developed at the Royal Radar Establishment (RRE), Malvern, enables pictures like these to be displayed in real time at remote sites using digitized weather radar data transmitted along telephone lines. Sequences of up to nine pictures can be displayed in quick succession to reveal the movement and development of rain areas.
(Crown Copyright, courtesy of the Controller, HMSO). (Used as cover of *Weather*, June 1974 (**29**, no. 6.))

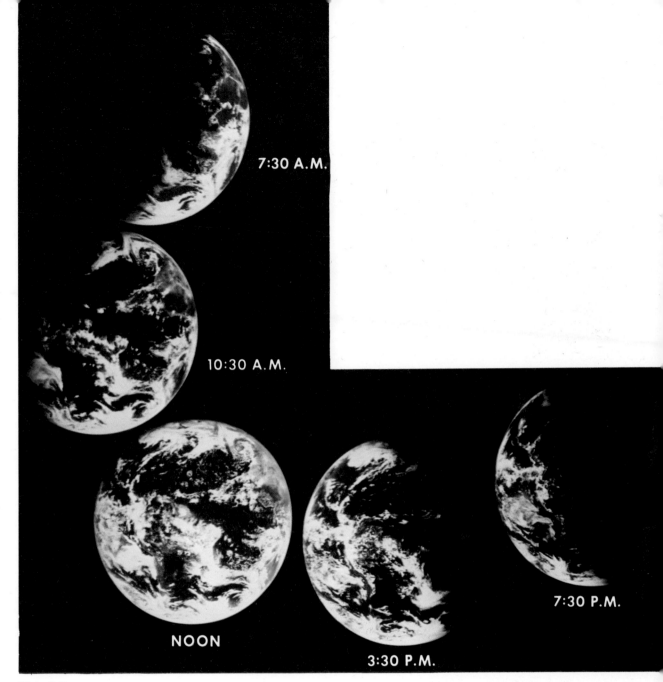

Plate 9.3
ATS geosynchronous satellite views of the illuminated portions of the disc of the Earth. ATS-III was in orbit above the mouths of the Amazone. These pictures show the passage of daylight from dawn to dusk around the globe. (Courtesy NASA)

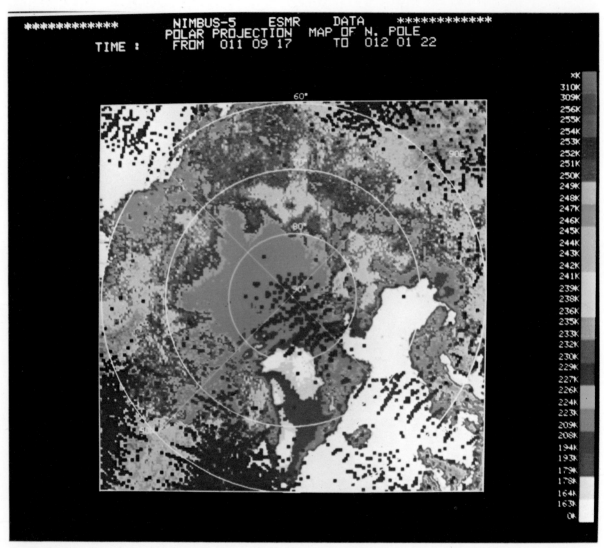

Plate 10.2

An ESMR thermographic map of the world for January 16th, 1973, obtained from Nimbus 5 in the 1.55 cm waveband, from an altitude of 1100 km. The maritime pattern is influenced by the general poleward gradient of temperature, patterns of ocean currents, and active rain areas. Over land the stronger microwave signals from the solid surface generally obliterate variations due to rain. (Courtesy, NASA)

Plate 14.3
At 1: 500 000 scale broad geographical patterns emerge. Geological features of the North Downs and farmland patterns of South East England together with the estuaries of the Thames, Crouch and Blackwater stand out in this processed image from Landsat 1 data.

Plate 14.2
Infrared colour photograph of part of the Cotswold scarp near Hawkesbury, Gloucestershire. Note harvested fields in blue-white strips, grass and ley fields in red. Some blue trees are diseased or dying. (July, 1972).
(Courtesy, Hunting Surveys Ltd., Boreham Wood, Herts.)

Plate 15.2
Infrared colour photograph of Durdham Downs, Bristol. Note the characteristic red imaging of healthy green vegetation. Diseased trees (including Elm) can be identified by their blue or grey colours. The underlying fold structures in Carboniferous Limestone image in light grey tones with patches of grass cover in red.
(Courtesy, Hunting Surveys Ltd., Boreham Wood, Herts.)

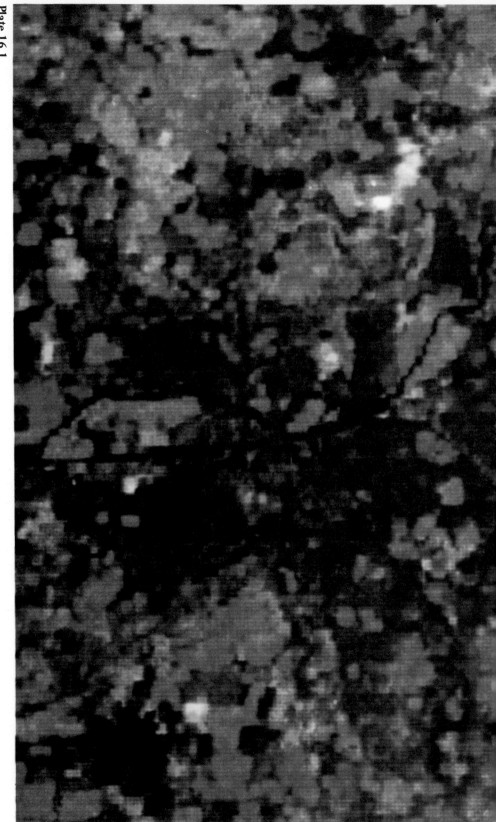

Plate 16.1
Computer processing of Landsat-1 data has produced this 1 : 50 000 image of Reading. Note the white/blue/
black renderings of urban features in contrast to adjoining farmland imaging in green, yellow and red.

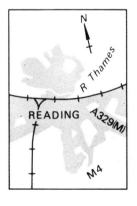

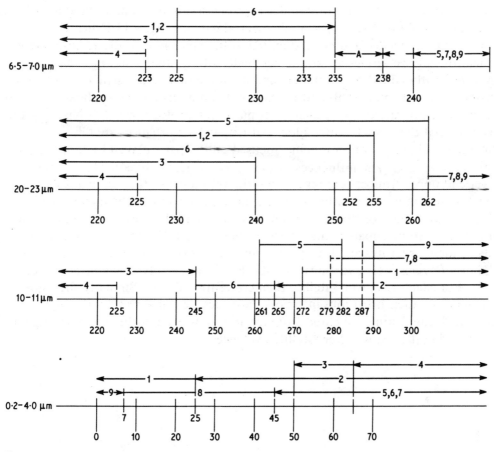

Key to cloud types:
1. Cirrus   2. Cirrus with lower clouds   3. Cumulonimbus and/or cirrostratus
4. Cumulonimbus   5. Middle clouds   6. Middle clouds with cirrus above
7. Stratus or stratocumulus   8. Cumulus   9. Clear   10. No decision

Fig. 9.11

A basis for nephanalysis using multispectral cloud data. This Cloud Type Decision Matrix was constructed for data retrieved from the Nimbus III instrument package. The horizontal graduations are K. Arrow heads indicates < or > Broken lines show the thresholds of nadir angles > 40°.
(Source: Shenk and Holub, 1973).

On the analytical front, efforts are being made to improve the nephanalysis procedure, particularly by automatic or 'objective' means. Three-dimensional nephanalyses have been compiled from SR infrared data (Plate 9.6), and work is proceeding here. A rather different possibility currently under investigation is that a new cloud identification tion procedure could be based upon multispectral data, as Fig. 9.11 indicates.

So far as our knowledge and understanding of the behaviour of the atmosphere is concerned, the biggest strides are still being made within the tropics. Fascinating studies have greatly added to our understanding of major weather structures like the inter-tropical cloud band (ITCB) and the South Asian monsoon. Of particular interest is the new 'scale-interaction' approach to tropical meteorology, which is based on the appreciation that the basic building blocks of tropical disturbances are relatively small and short-lived. Much attention is being focused on the parts played by 'hot-towers' of tropical convection. The influences of these, the principal dynamic links between the lower and upper troposphere range from the very local to the sub-hemispheric scale. Other areas in which weather satellites have much to offer the meteorologist include middle latitudes of the southern hemisphere, and the north and south polar regions. It has become apparent that southern hemispheric depressions differ in many respects of their structures and life-cycles from their northern hemispheric counterparts. Maps of cyclogenesis, vorticity distributions, and cyclolysis can be compiled much more satisfactorily from satellite than conventional data. Similarly the associated fronts, and their intrusions into polar latitudes (revealed well by the infrared imagery which can distinguish cloud from ice and snow even during the polar night) can be studied best from weather satellite evidence.

References

Anderson, R.K., and Veltischev, N.F. (1973), 'The use of satellite pictures in weather analysis and forecasting', *World Meterorological Organization, Technical Note No. 124,* W.M.O., Geneva.

Aracon: (1971), *The Best of Nimbus,* Contract No. NAS 5-10343, Allied Research Associates Inc., Concord, Massachusetts.

Barnett, J.J. *et al.* (1972), 'The first year of the selective chopper radiometer on Nimbus 4', *Q. J. R. Met. Soc.,* **98**, 17.

Barrett, E.C. (1970), 'Rethinking climatology', in *Progress in Geography, Vol. II,* Edward Arnold, London, pp. 154-205.

Barrett, E.C. (1974), *Climatology from Satellites,* Methuen, London, pp. 61-146.

Battan, L. (1973), *Radar Observation of the Atmosphere,* Chicago University Press, Chicago.

Hoppe, E.R., and Ruiz, A.L. (1974), 'Catalog of operational satellite products', NOAA Technical Memorandum, NESS 53, U.S. Department of Commerce, Washington.

Hubert, L.F., and Timchalk, A. (1969), 'Estimating hurricane wind speeds from satellite pictures', *Monthly Weather Review,* **97**, 383.

Schwalb, A. (1972), 'Modified version of the improved Tiros operational satellite (Itos D-G)', NOAA Technical Memorandum, NESS 35, U.S. Department of Commerce, Washington.

Shenk, W.E. and Holub, R.J. (1973), 'A multispectral cloud type identification method using Nimbus 3 MRIR measurements', presented to the Conference on Atmospheric Radiation, Fort Collins, Colorado.

Spalding, T.R. (1974), 'Satellite data for tropical weather forecasting', in *Environmental Remote Sensing applications and achievements,* Barrett, E.C. and Curtis, L.F. (eds.), Edward Arnold, London, pp. 215-240.

Suomi, V.E. (1970), 'Observing systems for weather "nowcasting" and "forecasting" , in *A Century of Weather Progress,* American Meteorological Society, Boston.

10 Global climatology

10.1 The nature of the problem

At the end of the previous chapter reference was made to atmospheric variations extending over months and years rather than hours or days. When we begin to think in terms of longer periods of time more smoothing and reduction of the shorter period data is usually necessary so that the more general trends may not be obscured by statistical noise. Studies of the atmosphere over periods upwards from about five days in length may be considered climatological rather than meterorological in nature and emphasis. As a consequence of the different time-scale involved the physical factors thought significant by climatologists may be different from those which are basic in meteorology. Although the advantages of remote sensing systems for weather studies (see p. 152) are all advantages for studies of climates also, the advantages of such systems for studies which are specifically climatological are somewhat different in emphasis. The chief of these advantages are as follows:

(a) Weather satellite data are much more nearly complete on a global scale than conventional data.

(b) Satellite data for broad, even global-scale areas, are more homogeneous than those collected from a much larger number of surface observatories.

(c) Such data are often spatially continuous, in sharp contradistinction to data from the open network of surface (point-recording) stations.

(d) Satellite observations are complementary to conventional observations and each may throw extra light upon the other.

(e) Satellites can provide more frequent observations of some parameters in certain regions.

(f) The data from satellites are all made by objective means (unlike some conventional observations, e.g. visibility and cloud cover) and are immediately amenable to computer processing.

However, it should not be thought that satellite data are, as a result of these advantages, an immediate, universal panacea to the problems of the climatologist. There are several new problems posed to the would-be climatological user of such data by their intrinsic characteristics. The chief of these problems include:

(a) The vast quantities of new data points involved.

(b) The physical indications of satellite data; these are often different from those of conventional observations.

(c) The selection and development of appropriate techniques to analyse the new types of information.

(d) The resolutions of the data. These are not always optimal for climatological uses.

(e) Degradations of components of the satellite—ground system complexes often complicate the evaluation and analysis of the data.

(f) It is often difficult to decide upon the best form for the presentation of results.

Notwithstanding the problems, many climatological products of interest and value have begun to appear since the runs of satellite data have lengthened sufficiently to make them possible. These include inventories of parameters as significant as the net radiation balance at the top of the Earth's atmosphere (the primary driving force of the Earth's atmospheric circulation patterns) and the distribution of cloud cover (a big influence on the albedo of the Earth/atmosphere system and its component parts, and an indicator of horizontal transport patterns of latent heat). In pre-satellite days certain components of the radiation balance (short-wave (reflected) and long-wave (absorbed and re-radiated) energy losses to space) were established only by estimation, not measurement. The only comprehensive maps of global cloudiness compiled in pre-satellite days depended heavily on indirect evidence, and could not be time-specific. So, satellites are affording us more comprehensive and more dynamic views of global climatology than were possible before their day (See Figs. 8.3(a)–(d)).

Satellites are also assisting the rise of applied macroclimatology. For example, active efforts are being made in several centres to develop computer models of the Earth's atmosphere, with a view to the improvement of methods of extended and long range weather forecasting. In such models the net radiation balance at the top of the atmosphere is one of the outputs. By comparing this with satellite-based net radiation patterns further insight is gained into the efficiency of the computer model, and those aspects of it which might be improved.

In the past, the development of climatology as a worthwhile and distinctive field of study has been singularly hampered by an inadequacy of data. Satellites are helping enormously to correct this chronic deficiency. Let us review in more detail the chief areas to benefit.

10.2 The Earth/Atmosphere Energy and Radiation Budgets

A distinction is sometimes drawn between:

(a) An energy budget or balance, which is the equilibrium known to exist when all sources of heat gain and loss for a given region or column of the atmosphere are taken into account. This balance includes advective and evaporative terms as well as a radiation term: that is to say both horizontal movements of energy, and the part played by absorptive gases in the atmosphere, are considered as well as the vertical radiation interchange.

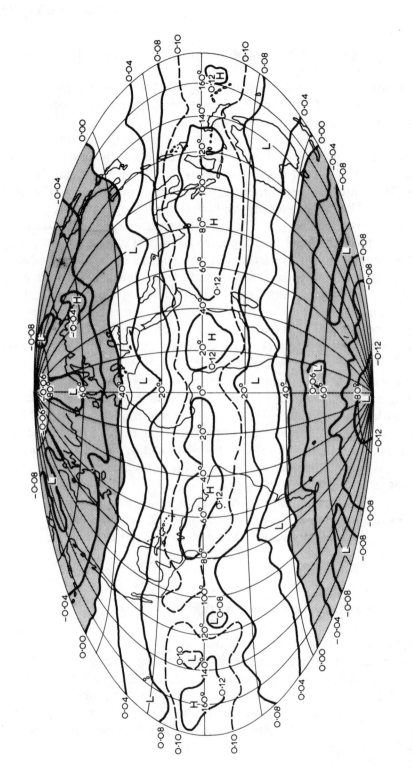

Fig. 10.1
Mean annual patterns of components of the Earth/atmosphere radiation budget ($RN_{EA} = I_0 - I_0 A - H_L$), 1962-5, from weather satellite evidence:
(a) Planetary net radiation balance (RN_{EA}) at the top of the Earth's atmosphere.

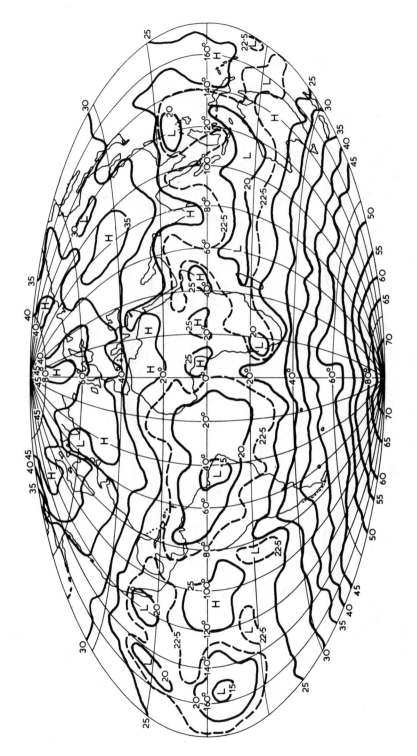

Fig. 10.1

Mean annual patterns of components of the Earth atmosphere radiation budget ($RN_{EA} = I_0 - I_0A - H_L$), 1962-5, from weather evidence: (b) Planetary albedo (I_0A) in percentages of incident radiation (I_0)

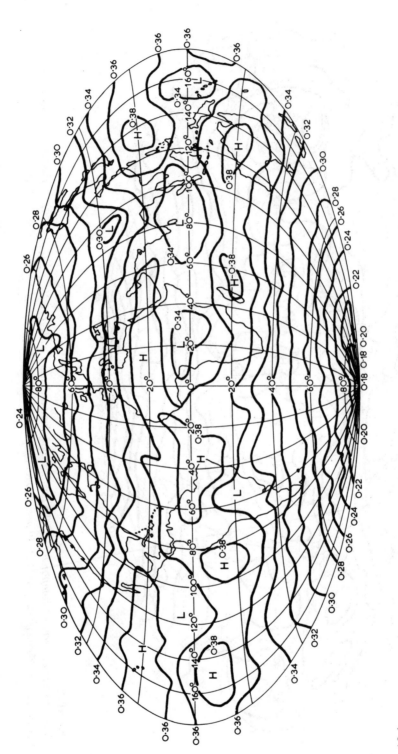

Fig. 10.1
Mean annual patterns of components of the Earth/atmosphere radiation budget (RN$_{EA}$ = I$_0$ −I$_0$ A−H$_L$), 1962-5, from weather satellite evidence (c) long-wave radiation (H$_L$)

(b) A radiation budget or balance. This is the equilibrium which exists between the radiation received by the Earth and its atmosphere from the Sun, and that emitted and reflected by the Earth and its atmosphere in return.

One outstanding problem is that, since the advective processes involved in the heat balance include movements both in the atmosphere and the Earth's oceans, these are still hard to evaluate from satellite data alone. Estimates are usually used instead, based on rates of atmospheric and ocean water flux, established by *in situ* sensors. However, much progress has been made in reassessing the Earth/atmosphere radiation balance from satellite data sources. The net radiation balance depends on three quantities, namely;

(a) The solar constant.

(b) The planetary albedo (the percentage ratio of that solar energy scattered and reflected by the atmosphere and Earth surface, to the total incident solar radiation. This depends largely upon the solar constant, the inclination of the Earth to the Sun's rays, and the reflecting capabilities of the Earth surface and its cloud cover).

Satellites are helping to evaluate (a) more accurately than before by measurements in the ultraviolet and visible wavebands. Infrared radiometers provide measurements of (b) and (c) through appropriate channels. Figs. 10.1 a,b, and c, illustrate mean annual geographical distributions of the parameters involved in the annual radiation budget. Table 10.1 integrates these parameters for global and hemispheric areas. The most significant conclusions from studies of these kinds include the following:

(a) Judged by one period of four to five years, the net radiation balance of the Earth (and either hemisphere separately) is in radiative balance. Thus there seems no requirement for an annual net energy exchange across the equator as assumed by many earlier workers.

(b) Both hemispheres are darker (mean albedo 30 per cent) and warmer (average long wave radiation 0.340 cal cm^{-2} min^{-1}) than the widely accepted estimates by J. London (1957) in presatellite days (35 per cent and 0.325 cal cm^{-2} min^{-1} respectively). The equivalent blackbody temperature difference is 3K (254K cf. 251K). If the Earth/atmosphere system is both warmer and darker than was previously thought, the system must accommodate — and probably transport — about 15 per cent more energy in each hemisphere.

(c) Each hemisphere has nearly the same planetary albedo and infrared loss to space on a mean annual basis. Since the surface features of the two hemispheres are quite different, clouds would seem to be the dominant influence on the energy exchange with space.

10.2.1 Regional details

Although some regional differences are apparent in Fig. 10.1 a-c, we need to view the world

Table 10.1
Mean annual and seasonal radiation budget data for the Earth/atmosphere system, from first generation weather satellites.

	Global average					Northern Hemisphere					Southern hemisphere				
	DJF	MAM	JJA	SON	Annual	DJF	MAM	JJA	SON	Annual	DJF	MAM	JJA	SON	Annual
I_0	0.51	0.50	0.49	0.50	0.50	0.34	0.56	0.65	0.42	0.50	0.69	0.43	0.32	0.58	0.50
H_a	0.34	0.35	0.37	0.36	0.35	0.24	0.39	0.48	0.31	0.36	0.46	0.30	0.25	0.41	0.35
H_r	0.16	0.15	0.12	0.14	0.15	0.10	0.18	0.17	0.12	0.14	0.22	0.13	0.07	0.17	0.15
A	0.31	0.31	0.25	0.28	0.29	0.29	0.31	0.26	0.27	0.28	0.32	0.30	0.22	0.29	0.29
H_L	0.32	0.33	0.33	0.34	0.33	0.32	0.33	0.34	0.34	0.33	0.33	0.32	0.32	0.34	0.33
$*RN_{EA}$	0.03	0.02	0.03	0.02	0.02	−0.07	0.06	0.13	−0.03	0.02	0.13	−0.02	−0.07	0.06	0.02

Where I_0 = incident solar radiation (cal cm^{-2} min^{-1})
    H_a = absorbed solar radiation (cal cm^{-2} min^{-1})
    H_r = reflected solar radiation (cal cm^{-2} min^{-1})
    A = planetary albedo (per cent)
    H_L = emitted infrared radiation (cal cm^{-2} min^{-1})
    RN_{EA} = net radiation budget of the earth/atmosphere system (cal cm^{-2} min^{-1})
* Probable absolute error of ± 0.01 cal cm^{-2} min^{-1}

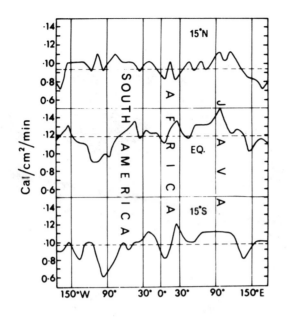

Fig. 10.2
Longitudinal variations in net inputs of energy (RN$_{EA}$) (mean annual case), evaluated from five years of satellite data. (Source: Vonder Haar and Suomi, 1971)

at a larger scale if more detailed pictures are required. Fig. 10.2 portrays the variation of net radiation with longitude for three latitude circles. It is clear that there are some variations related to the contrast between land and sea, and others which seem to be related to air mass and circulation differences.

The influence of geography on the net radiation balance is illustrated most vividly by Figs 10.3a and b. The radiation balance in middle and high latitudes of the northern hemisphere in a summer month is seen to assume a highly fragmented pattern. Some areas – even within the Arctic Circle – experience a net radiation gain, while the chief area of net radiation loss is over the Greenland ice-cap, neither over, nor very near, the pole. Generally the areas of net gain ('sources' of radiation energy for the general circulation) indicate little cloud cover, and the areas of net loss (the radiation 'sinks') indicate high percentages of cloud, or, in the Arctic Ocean, permanently frozen surfaces.

In the same period winter held the south polar region strongly in its grip. Fig. 10.3b shows the whole area south of 40° S to be in deficit, though, curiously at first sight, the deepest radiation sink is not over Antarctica itself, but over the surrounding Southern Oceans. This is the zone in which most drastic environmental changes take place from one high season to the next. Most of Antarctica is always frozen, and radiation losses in winter are relatively small since there is little stored or advected energy available to be emitted as long-wave radiation (heat energy) to space. Meanwhile, the surrounding

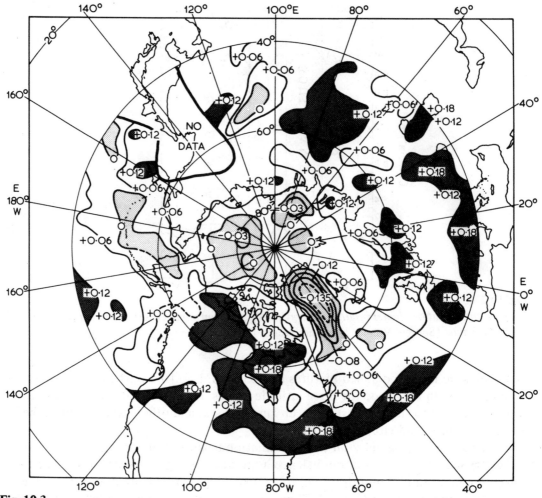

Fig. 10.3

(a) Net radiation balance (RN_{EA}) of the Earth/atmosphere system over the Arctic based on Nimbus II data, July 1st-15th, 1966 (cal cm^{-2} min). Areas in surplus are strippled. (Source: Barrett, 1974)

oceans freeze in winter, and much energy stored through the summer months is lost as long wave radiation accompanying the phase change of water from liquid to solid. From the meteorological point of view, once frozen, the oceans act more like continents. In particular, their albedoes suddenly and dramatically increase, and the difference received radiation and that lost by scattering and reflection tilts the net radiation budget to its winter level of considerable deficit.

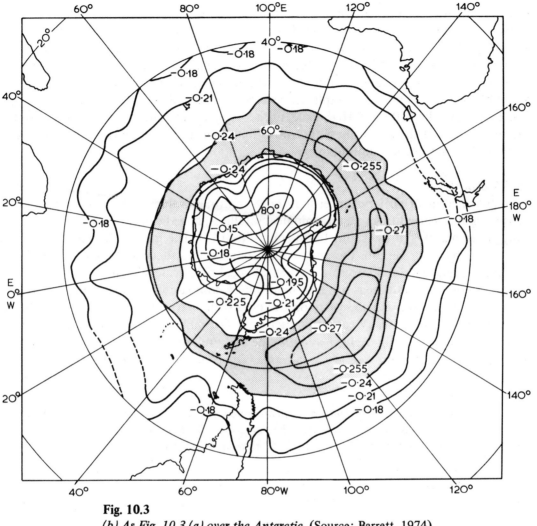

Fig. 10.3

(b) As Fig. 10.3 (a) over the Antarctic. (Source: Barrett, 1974)

10.3 Global moisture distributions in the atmosphere

10.3.1 Water vapour

This gas, so important for the absorption and transport of energy — and as the raw material from which clouds and precipitation are fashioned — is rather hard to measure and map by remote sensing means. In the thermal emission wavebands of the infrared region of the electromagnetic spectrum radiation in the water vapour absorption band from 5.7–6.9 μm emanates mostly from water vapour when clouds are absent, but partly from within the upper layers of clouds when these are present. Consequently both the cloud cover in each column and the temperatures of any cloud tops must be known if contaminating contributions from such sources are to be removed from water vapour waveband radiances. Data from MRIR and HRIR radiometers have been used in a rather restricted way for mapping global water vapour mass or relative humidity. Such studies have usually concentrated on the upper troposphere (above 500 mb). Below 500 mb the complications caused by clouds are considerable. Spectrometers like SIRS and IRIS have, in theory, the capability to profile the water vapour content of the atmosphere, but no climatological (time-averaged) results based on their data have been published yet.

10.3.2 Clouds

We mentioned in an earlier chapter that daily brightness values, derived from visible image signals received from meterorological satellites, have been processed to give daily cloud pictures of the sunlit Earth (pp. 140–143). The mesoscale archive has been used as the basis for perhaps the most comprehensive and successful approach to the mapping of relative cloud cover on a climatological time scale. The mesoscale matrix is comprised of 512^2 unit areas for each hemisphere compared with the 4096^2 points in the full resolution matrix. A grid square on the mesoscale matrix is one sixty-fourth the area of an American Numerical Weather Prediction — Global Weather Center grid square. Using a set of empirically-derived weights, the daily relative cloud cover may be estimated in oktas (eights) for each mesoscale area (see Table 10.2). By saving the daily

Table 10.2

Empirically derived weights for estimating the daily relative cloud cover. These were selected to achieve the best agreement in cloud amounts between manual (visual) estimated and automated estimates.
(Source: Miller, 1971)

Original brightness range	Class	Contribution to total cloud amount %	Weights October-May	June-September
0, 1, 2	1	0	0	0
3, 4, 5	2	25	2	2.5*
6, 7, 8	3	88	7	7.5*
9, 10, 11	4	100	8	8
12, 13, 14	5	100	8	8

* The summer and winter weights were applied to the Northern and Southern hemispheres separately according to the hemisphere seasons. It was noted that the higher frequency of occurrence of small cumulus cloudiness in the summer hemisphere warranted the higher weighting during the summer months.

values by the month for the entire period of the record, these values may be grouped in a 10-class frequency distribution including 0-8 oktas of cloud cover, plus one class (9 'oktas') for missing data. A range of mean ('relative') cloud cover maps has been prepared for 1967-70, as illustrated by Plates 10.1a-d.

Several satellite-based methods have had the aim of producing mean cloud maps or other forms of time-averaged spatial displays. In the first decade of weather satellite operations the most popular techniques were based on nephanalyses, using their generalized indications of the percentage of sky which is cloud covered. Such techniques are based on sets of weighting factors through which the nephanalysis categories of cloud cover (commonly C, MCO, MOP, O, and Clear) may be translated into acceptable cloud percentages. The techniques differ mostly in their sampling procedures. Some workers based their mean cloud maps on the nephanalysis indications at selected intersects of latitude and longitude, completing them by interpolating isonephs. Others have based their maps on estimates of the mean cloud cover in each grid square of given size. These maps have been presented either in choropleth (area shaded) or isopleth form, assuming in the latter case that the mean cloud cover in each square may be taken as representative of its centre point.

Any maps prepared from nephanalyses must be subject to errors arising from their own degrees of generalization, as well as those implicit in the nephanalyses themselves. The most obvious shortcomings of nephanalyses stem from the need for subjective judgements to be made by the nephanalysis compilers. It is widely accepted that nephanalysis-based mean cloud maps tend to over-estimate cloud cover at the upper end of the range, and to under-estimate it at the lower end. This is because the satellite imaging systems, and the nephanalysts, may eliminate small breaks in cloud fields where the proportion of cloud-covered sky is high, and fail to resolve and represent small cloud units where the sky is mostly clear. Despite these problems the patterns revealed by hand analyses of mean cloud cover have been generally acceptable, and of considerable interest where new climatological features (like the split ITCB, for example) have been revealed. On the credit side, such mean cloud maps are contaminated less by the 'background brightness' of surface phenomena which may be mistaken for clouds in automated ('objective') procedures.

One early objective technique for mean cloud mapping achieved some success in the late 1960's. This involved the production of 'multiple exposure average' pictures for a selected major region of the world for a chosen period of time. A photographic plate was exposed to each of a number of daily computer-rectified, full resolution picture products (4096^2 picture points) in turn. Multiple exposure averages, whilst being interesting climatological statements, suffer from two inherent characteristics. First, they do not differentiate between brightness due all or mainly to highly reflective Earth surfaces as distinct from that due to clouds. Second, they are qualitative, not quantitative displays. More recently, efforts have been made to develop techniques which lack such deficiencies.

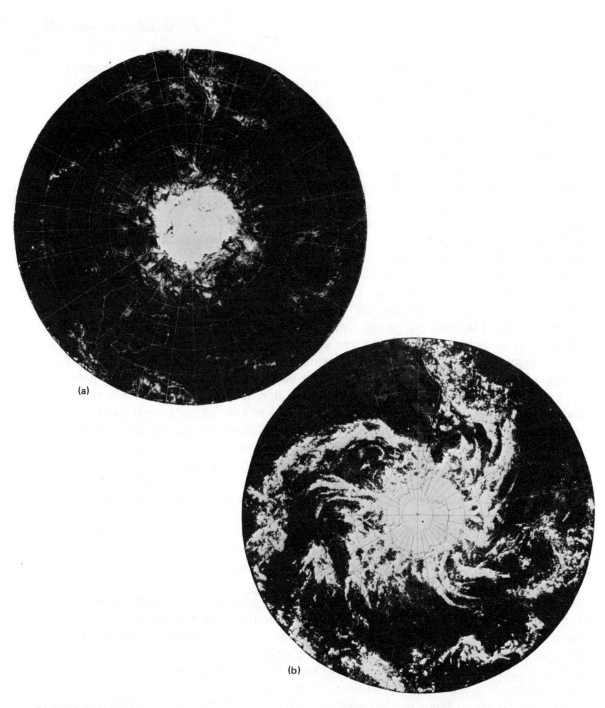

Plate 10.1
Brightness composite displays from Essa weather satellite photography: (a) 5-day minimum brightness composite; (b) 5-day maximum brightness composite; (c) 30-day average brightness composite; (d) 90-day average brightness composite. (Courtesy, ESSA)

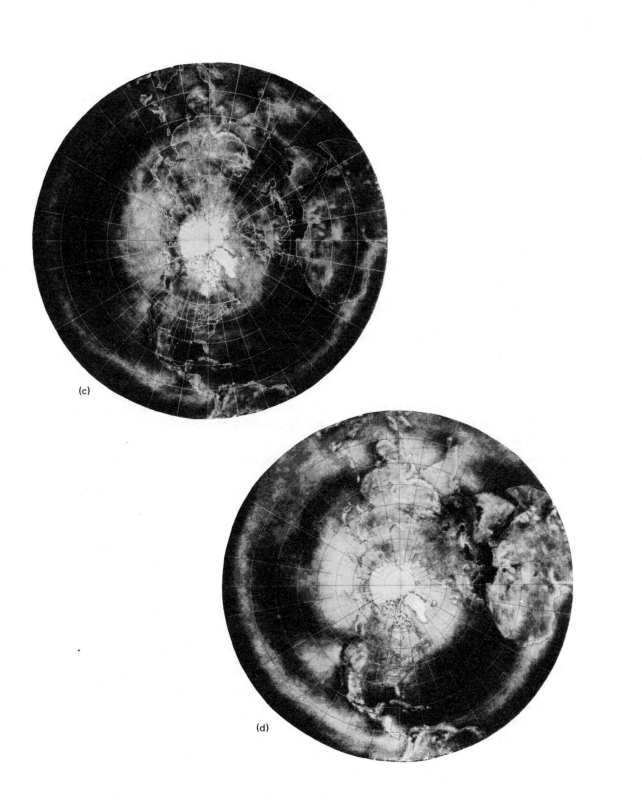

(c)

(d)

Comparisons between picture brightness values (converted to 'relative cloud cover') and conventional observations have been made by staff at the USAF Environmental Technical Applications Center. The studies have revealed gross differences between point observations of total cloud amount made at the surface, and the satellite cloud cover estimates: the satellite and surface observations have a similarity in pattern, but point-to-point values often differ. A general tendency is for the satellite data to suggest lower cloud amounts than concurrent surface estimates. These differences can be ascribed primarily to differences between the field of view of the surface observer and the angle of view from the satellite. To some extent the lower response threshold and resolving capability of the satellite sensor may also be important. Unfortunately, 'background brightness' is not extracted in the present archival programme, and strongly-reflective surfaces such as ice, sand and snow contaminate the 'relative' cloud cover' maps. In future it may be possible to remove persistent brightness associated with the surface of the Earth, leaving the more mobile brightness caused by clouds. This has already been achieved in specialized studies, for example ice margin 'mapping' as described on p. 219.

10.3.3 Rainfall

The possibility of mapping rainfall distribution from satellite data is receiving close attention in several parts of the world. Strictly speaking there are two rather different aspects to this problem, namely:
(a) Mapping rainfall for short-term weather and/or hydrological forecasting purposes.
(b) Mapping rainfall for climatology.

We need not differentiate between these in this brief review. Instead, let us consider very broadly the types of approaches which are being followed. Some will receive closer attention in the next chapter.
(a) Statistical techniques. It has been demonstrated that an areal statistics approach can yield worthwhile results on both a daily and a monthly basis. Satellite-viewed clouds are ascribed indices relating to the probabilities and intensities of associated rain. These can be correlated with rainfall observations by regression techniques. Weighting factors are calculated to translate new observations of cloud cover in forecasting situations into rainfall estimates. Satellite estimates for data-sparse areas may be combined with gauge measurements for the best results.

(b) Cloud brightness techniques. These are based on the empirical fact that actively precipitating clouds are often brighter than the rest. Unfortunately this difference is more useful for making 'rain versus no rain' decisions than quantitative estimates of time-integrated rainfall totals: the relationship between cloud brightness and rate of rainfall is very variable. One reason is because the brightness of clouds increases rapidly with cloud thickness, reaching saturation level as early as clouds 600 m thick. Even such dissimilar clouds as thick cirrostratus and towering cumulonimbus may be hard to differentiate on the basis of brightness alone.

(c) Infrared techniques. Detailed studies employing Nimbus II HRIR data have been based on the assumption that (at least in the tropics) the intensity of convective activity is related to the heights of cloud tops. It seems possible that cloud-height maps might usefully be compared with precipitation maps compiled from radar or ground observations so that eventually the cloud top maps might be used alone as indicators of rates and totals of associated rainfall.

(d) Parameterization techniques. These consider the physics of convective clouds. In pilot studies for the tropics the satellite contribution is the percentage of each unit grid square which is covered by convective cloud.

(e) Active microwave techniques. Radar observations can be correlated with conventional data so that weighting factors may be evaluated for the estimation of rainfall from radar indications alone. (See Chapter 11).

(f) Passive microwave techniques. Data from passive microwave radiometers such as ESMR (Electrically Scanning Microwave Radiometer) on Nimbus 5 can be processed to give maps of precipitation intensity in mm h^{-1}. Unfortunately, the strength of the background signal over land in the 19.35 GHz (1.55 cm) region of the electromagnetic spectrum restricts the immediate applicability of such techniques to ocean areas, and the resolution is rather low. (Plate 10.2, see colour plates).

We may conclude that no one method has been generally accepted yet. Methods for rainfall estimation will continue to be developed, refined and validated for some time to come. It is probable that different approaches will be required for different latitudes. Methods based on geosynchronous satellite data (most immediately methods (b), (c) and (d) above will be best for the tropics, whilst method based on the less-frequent lower-altitude satellite data (like methods (a) and (e) above) will be required for middle to high latitudes.

10.4 Wind flows and air circulations

Although there are several techniques for assessing the speed and direction of wind flow from satellite evidence, these have been utilized almost entirely in meteorological, not climatological, contexts. For example, estimates of maximum wind speeds in hurricanes are made routinely by the U.S. Weather Bureau. Similarly, the directions of wind flow can be established better in some circumstances from satellite cloud images than conventional data, but the contributions of satellite-derived winds to climatological statistics have, so far, been too few and sporadic to be of great significance. Satellite observations for air flow analyses have their greatest potential in the tropics, where geosynchronous data may yield a variety of wind and circulation parameters. (See Chapter 9, p. 170). Already some daily wind charts for tropical regions have been corrected using wind speed and direction data from ATS satellites. Since mean monthly, seasonal and annual maps of isotachs and streamlines are prepared from such

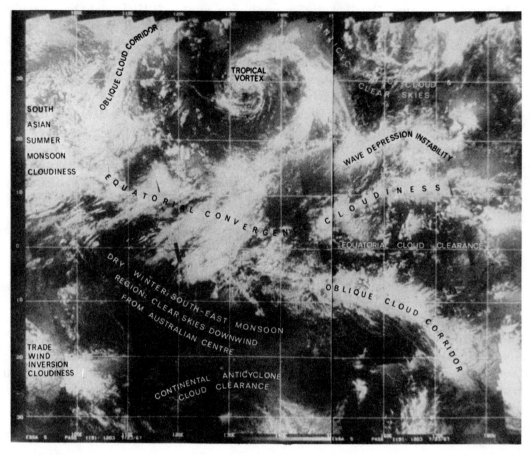

Plate 10.3
A tropical vortex and a wave pattern over the western tropical North Pacific; July 23rd, 1967.
(Source: Barrett, 1971).

sources to summarize the circulation patterns, we can say that satellite data are beginning to contribute to climatological statements of this type too, though less completely than in the case of radiation maps or even rainfall distributions.

10.5 The climatology of synoptic weather systems

Remembering that many atmospheric weather systems are clearly distinguishable in weather satellite data products especially through their attendant cloud and radiation temperature fields, it is not surprising that many new climatological facts have

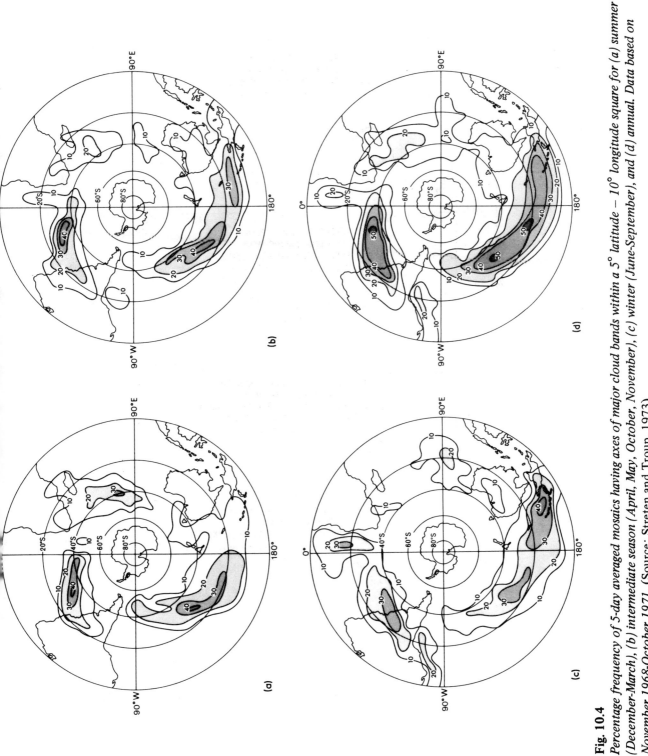

Fig. 10.4
Percentage frequency of 5-day averaged mosaics having axes of major cloud bands within a 5° latitude – 10° longitude square for (a) summer (December–March), (b) intermediate season (April, May, October, November), (c) winter (June–September), and (d) annual. Data based on November 1968–October 1971. (Source: Streten and Troup, 1973)

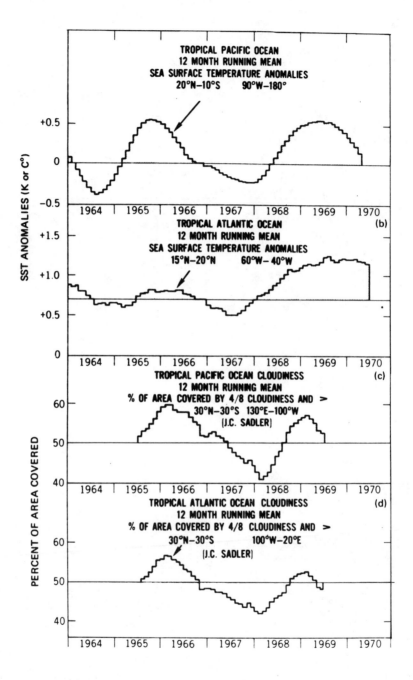

Fig. 10.5

A comparison of 12-month running means of sea surface temperature anomalies and satellite-derived cloudiness in percent of area covered by ≥ 4/8 cloudiness in specified sections of the tropical Pacific. (Source: Allison *et al.*, 1972)

emerged from studies of their time frequencies, and spatial distributions through extended periods. Perhaps the most interesting of such studies have been carried out for relatively inaccessible regions, though worthwhile results have also emerged for the better known land areas, especially in the realms of jet stream climatology.

Particular light has been thrown upon the structure of the intertropical cloud band (ITCB), a prominent feature on visible and infrared images of the tropics. The position, range of forms, and secular shifts of the ITCB have been elucidated around the globe, and regional and seasonal contrasts described. Lateral to the ITCB, wave-like cloud features, associated with pertubations in pressure and wind fields, have been identified not only in the tropical North Atlantic (where easterly waves were first described) but in most if not all of the other tropical ocean regions too. (Plate 10.3). The climatology of tropical revolving storms (especially hurricanes) has been much improved, and that of mid-latitude vortices and frontal bands especially in the southern hemisphere (Fig. 10.4). Currently, much interest is being shown in studies of atmospheric cycles in different parts of the world (Fig. 10.5), the possible relations between them in terms of the strengths and frequencies of synoptic weather systems ('atmospheric teleconnections'), and the possible applications of such knowledge and understanding in climatic prediction. At last we are able to view the whole atmosphere either as a single, complex fluid continuum, or as a sum of many inter-related, and more or less interdependent parts. The 1970's are particularly exciting years for the climatologist.

References

Allison, L.J. *et al.,* (1972), 'Air-sea interaction in the tropical pacific Ocean', NASA Technical Note NASA TN D-6684.

Barrett, E.C. (1974), 'Tropical climatology from satellites', *J. Brit. Interplanetary Soc.,* 27, 48.

Barrett, E.C. (1974), *Climatology from Satellites,* Methuen, London, pp. 147-313.

Barrett, E.C. (1971), 'The tropical Far East: high season climatic patterns derived from ESSA satellite images', *Geographical J.* 137, 535.

Martin, D.W. and Scherer, W.D. (1973), 'Review of satellite rainfall estimation methods', *Bull. Amer. Met. Soc.,* 54, 661.

Miller, D.B. (1971), 'Global atlas of relative cloud cover, 1967-70, based on data from meteorological satellites', U.S. Air Force, U.S. Department of Commerce, Washington.

Streten, N.A. (1973), 'Some characteristics of satellite observed bands of persistent cloudiness over the Southern Ocean', *Monthly Weather Review,* 101, 486.

Streten, N.A. and Troup, A.J. (1973), 'A synoptic climatology of satellite-observed cloud vortices over the southern hemisphere', *Q. J. Roy. Met. Soc.,* 99, 56.

Vonder Haar, T.H. and Suomi, V.E. (1971), 'Measurements of the Earth's radiation budget from satellites during a five year period, Part I: extended time and space means', *J. Atmos. Sci.,* 28, 305.

11 Water in the environment

11.1 The importance of water

Water is the most ubiquitous yet most variable of the mineral resources of the world. Alone among the constituents of man's environment water may be found simultaneously in one area as a liquid, a gas, and a solid. Unlike most other Earth resources, water (on land and in the atmosphere) is continuously variable in its availability in one state or another. A further complication is that, geographically speaking, its patterns of concentration are always changing.

Apart from the purely scientific interest water arouses, man has a keen practical interest in its behaviour and distribution on account of its significance as a basic requirement of living organisms, and the consequent dependence of man himself and his economy upon it.

For these reasons — both academic and applied — the satisfactory monitoring of environmental water is one of the most difficult yet pressing problems confronting remote sensing as a distinctive discipline.

The hydrological cycle of the Earth/atmosphere system can be represented in simplified form as shown in Fig. 11.1. This cycle is comprised of three types of com-

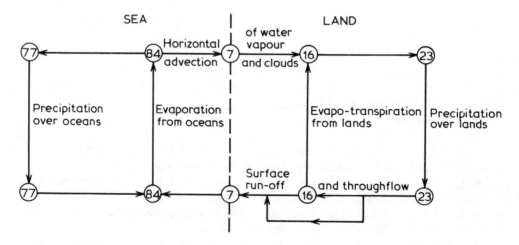

Fig. 11.1
A summary of the global hydrological cycle.

ponents namely storages, transports, and exchange processes. By far the greatest reservoir of water is the oceanic girdle of the globe, although significant amounts of water are retained temporarily within the rocks of the Earth's crust, in ice caps and snow fields on the surface of the Earth, in soils and regoliths, in fresh water reservoirs like lakes and inland seas, and in the atmosphere. Water is transported by circulations in the atmosphere and oceans, and by movements over and near the surfaces of land areas chiefly under the influence of gravity. It is left to the exchange processes to effect transformations of water from one physical state (solid, liquid or gas) into another. Since questions concerning water stored in, and transported by, the atmosphere have been discussed already in the chapters on meteorology and climatology, we may concentrate our attention here on continental and oceanic water, beginning with the exchange process by which atmospheric water becomes surface water, namely precipitation. Microwave systems offer the most immediate hope for obtaining data on rainfall intensity as well as distribution.

11.2 Precipitation measurements

11.2.1 Rainfall estimation using active microwave systems
In any review of the hydrological cycle a convenient starting point is the water supplied to land and sea surfaces by the exchange process of precipitation. This can be assessed in greatest detail by ground-based remote sensors of the active variety. In particular, microwave sensing systems of the radar family (see Chapter 9), have two important applications in the field of rainfall estimation:
(a) Radar systems can provide real-time estimates of rainfall intensities over selected areas. For certain problems, such as river management and the control of soil erosion, the short-period rainfall intensity is an important factor.
(b) Radar systems can give estimates of the total rain which has fallen over, say, a basin or a catchment through a specified period of time. Such information is often vital to water management programmes generally, and predictions of the likely flows of streams and rivers in particular.

By established practice such rainfall data are conventionally obtained from networks of rain gauges, preferably of a continuous-recording type. However, it is common for the rainfall stations to be quite widely spaced. Especially when rain falls from convective clouds, e.g. thunderstorm clouds, the rain gauge data may be very misleading so far as average intensities of rainfall across wide areas are concerned. Dependent depth/area (volume) estimates of rainfall through periods of time may be quite inaccurate. Shortly after World War II it was recognized that radar was capable of observing the location and areal extent of rain storms, and that, in conjunction with some rain gauge data, more accurate maps of rainfall might be obtained from data from such a source.

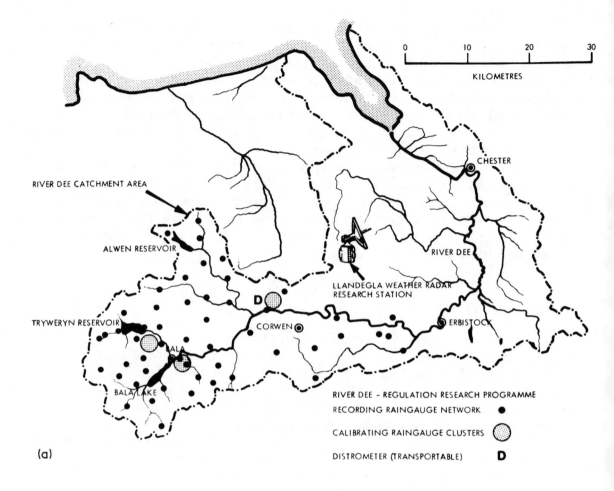

(a)

Fig. 11.2
The Chester Dee weather radar project. (a) Catchment instrumentation.

If we know the technical specification and performance of a radar set, the radar equation basic to most rainfall studies may be written in simplified form as follows:

$$\bar{P}_r = \frac{cZ}{r^2}$$ (11.1)

where $\bar{P}_r$ is the mean received power (the 'signal') from the target, c is a constant (the speed of light), Z a reflectivity factor relating to the type of precipitation in the target area, (Plate 11.1), and r the range (or distance) between the radar and the reflecting shower of rain. The reflectivity factor is of particular significance whether the output required is rainfall intensity or amount. Z is high where the raindrops are large and

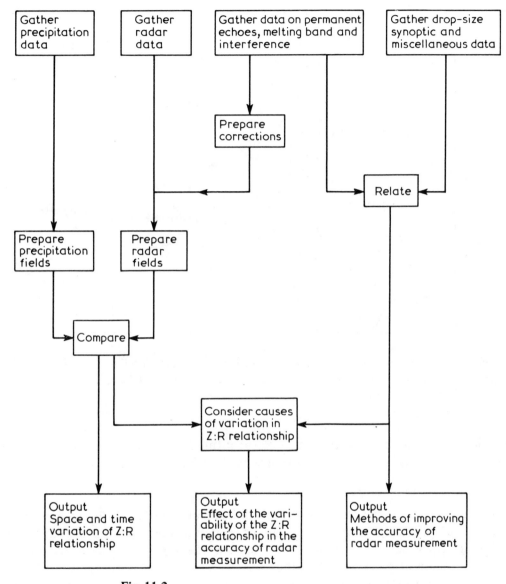

Fig. 11.2
The Chester Dee weather radar project. (b) Activity diagram.

numerous, and low where they are small and few. Quantitative estimates of rainfall (R) may be derived from the target echoes by evaluating the relationship $Z = aR^b$, where a and b are local constants established by empirical means. In middle latitudes, typical solutions for different types of rain are $200R^{1.71}$ for orographic rain, and $486R^{1.37}$ for

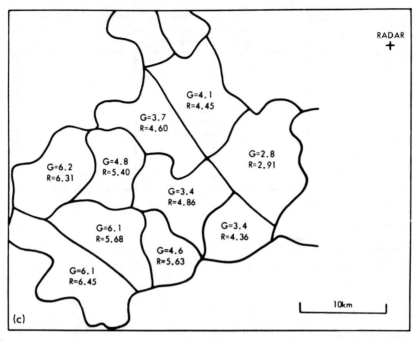

Fig. 11.2
The Chester Dee weather radar project. (c) Representative results: rainfall totals (mm) derived from radar (R), and raingauge (G), over sub-catchments for the period 1530-1630. Nov. 7th, 1971.
(Source: Plessey Radar, 1971)

thunderstorm rain. Such relations may be substituted for Z in Equation 12.1 so that, for operational usage, the radar equation (in simplified form) becomes:

$$\overline{P}_r = \text{constant} \times \frac{R^b}{r^2} \tag{11.2}$$

The implication is that for different latitudinal zones and for different types of rain the radar signal is proportional to the rate of rainfall divided by the square of the range.

The principles have been employed by many workers on both sides of the North Atlantic. A current project involves the Chester Dee in North Wales. This is being undertaken co-operatively by the U.K. Meteorological Office, the Water Resources Board, the Dee and Clwyd River Authority, and Plessey Radar. The area covered measures approximately 1000 km². This has been carefully instrumented to facilitate the necessary statistical comparisons between radar and conventional parameters (see Fig. 11.2). Especially in such areas of accidented relief it is necessary to establish the patterns of permanent, topographically-induced, echoes upon which rainfall echoes may be superimposed. The sequence of steps in the rainfall estimation process is set out in Fig. 11.2b. Sample results for a particular rainstorm are illustrated by Fig. 11.2c. A good measure of similarity exists between the two. Unfortunately in very mountainous, inaccessible, terrain — the type of area from which radar estimates of rainfall might be most valuable — the narrow beam, low elevation type of radar currently used in work of this kind would be largely ineffective. Were a higher elevation beam to be used instead the distance over which the rainfall estimates were valid would be relatively short, because ever higher regions of the atmosphere are sampled as range is increased. As a consequence, where there are large and important watersheds in mountainous regions, present systems and techniques can only cover a small area by each radar set. Perhaps new techniques — possibly involving a form of Doppler radar — can be devised to deal with this problem. Certainly for many hydrological purposes it is important to know as accurately as possible the scale, and variation through time, of this principal input into surface water hydrology.

11.2.2 Rainfall estimation using passive microwave systems

Mention was made in Chapter 10 (p. 196) of the new ESMR passive microwave system flown for the first time on Nimbus 5. It is worth adding some further comments on the potential of the ESMR system for rainfall mapping, for the areal capability of such a system is much greater than that of a single radar set. It is unfortunate that this potential of ESMR is restricted over land. Here the emissivity in the waveband in question is very high (≈ 0.9) compared with emissivities between $0.3-0.5$ over water surfaces. Furthermore, even over oceans the brightness temperature cannot be interpreted unambiguously in terms of any one atmospheric variable. Rather it is a linear combination of surface wind, water vapour, and liquid water content. However, in rain areas the hydrometeors are larger than in the absence of rain and resonant effects greatly increase the absorptivity of the clouds. In practice (see Plate 11.2) raining areas in clouds are effectively opaque at frequencies of 19.35 GHz. They appear as the darkest areas in the microwave images. The data received from ESMR are affected by water vapour, clouds, rain, soil moisture and snow and ice cover; information on all these hydrologically-significant parameters is available, but rarely without ambiguity. Additional data and interpretation skill are required to resolve the ambiguity.

11.3 The hydrology of land surfaces

11.3.1 Non-satellite studies

The problem at the heart of studying the hydrology of land surfaces by remote sensing is the high degree of variation — both temporal and spatial — which is exhibited by most hydrological variables. As a consequence questions of levels of precision and frequency of measurement are especially significant in the planning of remote sensing programmes in hydrology. In small-scale studies, for example of local catchments draining into a

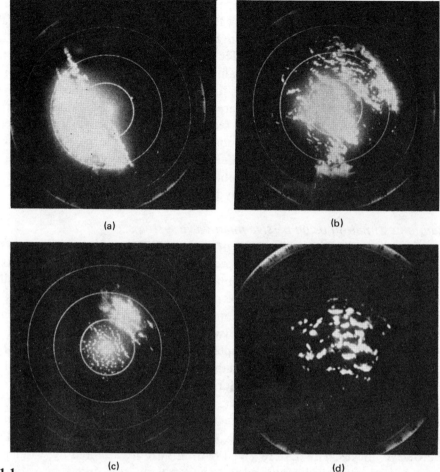

(a) (b)

(c) (d)

Plate 11.1

Typical maritime precipitation radar patterns interpreted in terms of the cloud characteristics with which they are likely to be associated. (PPI displays, 150 mile range). (a) Continuous stratiform cloud; (b) Ragged stratiform patches; (c) Small convective cells; (d) Large convective cells. (Source: Nagle and Serebreny, 1962)

dam or reservoir, the relationships which may be investigated include those depicted by Fig. 11.3. The related observations which seem to lend themselves most readily to evaluation by remote sensing means are listed in Table 11.1. It would appear that remote sensing systems may make up in increased frequency of observation what they currently fail to achieve in precision. In other words, a loss of precision in single-point measurement can be accepted in exchange for a greatly increased frequency of coverage over

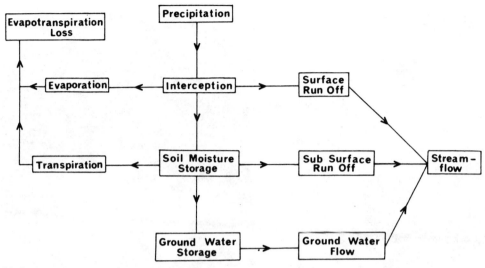

Fig. 11.3

A water-balance schematic for individual catchments. Certain of the components lend themselves more readily to assessment by remote sensing than others (See Table 12.1). (Source: Painter, 1974)

Table 11.1

Requirements of remote sensing systems for monitoring hydrological variables. (Source: Painter, 1974).

Observation	Ground instrument precision	Remote sensing requirement Precision	Frequency
Albedo	2–3%	10%	daily
Surface temperature	$\frac{1}{10}°C$	$\frac{1}{2}°C$	6-hourly
River depth	1 mm	50 mm	3-24-hourly
Point precipitation	1 mm	5 mm	daily
Point snow depth	5 mm	75 mm	daily
Water equivalent snow	1-2 mm	5 mm	daily
Suspended sediment concentration	1 mg^{-1}	20 mg^{-1}	daily
Soil moisture volume fraction	10%	20%	daily-weekly

a broad area. Thus potentially, the most useful applications of remote sensing in hydrology are likely to be those where spatial variations are more important than temporal variations.

The measurements which seem likely to benefit most from ground-based or airborne remote sensors include snow depth, soil moisture, and some variables necessary for computations of evaporation and evapotranspiration. The techniques which have been tested include black-and-white, colour, infrared and multispectral photo-

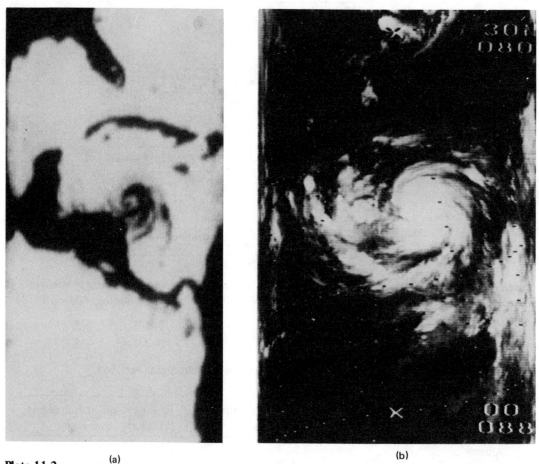

Plate 11.2

Hurricane Fifi, September 18th, 1974, as seen by Nimbus 5 through its Electrically Scanning Microwave Radiometer (a), and its Temperature Humidity Infrared Radiometer (b). The former reveals rain areas embedded in the cloud vortex so prominent in the latter. (Courtesy, NASA)

graphy, stereoscopic photography, and measurements of natural radioactivity from the soil by nuclear counting devices. The platforms which have been used range from surface platforms viewing distant slopes, to towers, tethered balloons and aircraft.

Apart from conventional photographic studies some of the most promising work that has been carried out to date has been based on aircraft-borne side-looking radar (SLR). One such project, codenamed RAMP (Radar Mapping of Panama) was initiated in 1965 by the U.S. Army Engineer Topographic Laboratories to demonstrate the peformance of high-resolution radar imagery in lieu of optical photography in heavily clouded regions. The sensing system was a K-band radar (see Fig. 2.2, p. 15). This has better than a 99 per cent cloud penetration capability, although it does not penetrate heavy rain or vegetative cover. Once-over coverage of the eastern end of the Republic of Panama was achieved in approximately four hours of flying time. A number of interpretational overlaps were prepared from the final mosaic. These included surface drainage patterns, and types of surface drainage regions (Plate 11.3 and Figs 11.4a and 11.4b) as well as general surface configuration (topographic provinces), vegetation and engineering geology.

It was possible to identify variations in the local and regional expression of surface drainage, for example differences in the density of streams and other drainage channels, and the depth of channel incisions. In turn such variations in the drainage patterns are diagnostic of specific terrain conditions, for example the nature and depth of soil cover, lithology, morphology, and the prevailing structural and tectonic influences. The patterns present themselves in an infinite variation of density and habit (or form). The density is determined mainly by the lithologic character of the rock traversed — itself a matter of considerable interest to hydrologists — involving its hardness, porosity, solubility and consolidation. The form relates mainly to structure, involving faults, fractures and the

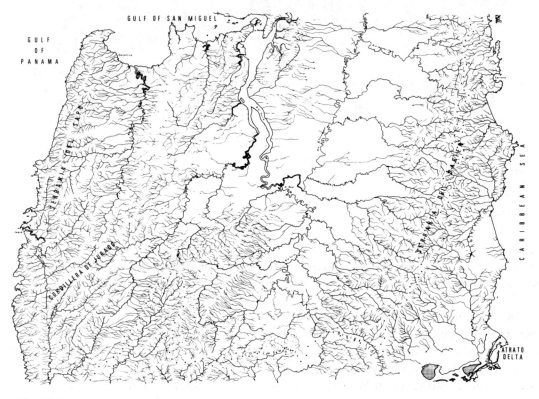

Fig. 11.4

Hydrologically-significant distributions mapped from a radar mosaic of the Darien region of eastern Panama, and northwest Columbia: (a) Surface drainage network;

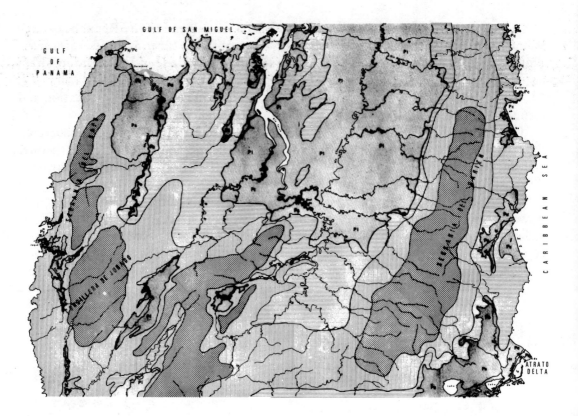

Regional Features

Plains — Relief generally less than 50m

Low hills — Relief generally 50–150m

High hills — Relief generally 150–600m

Mountains — Relief generally more than 600m

Local Geomorphic Features

Pu—Upland plain
Pc—Coastal plain
Pi—Interior plain
Pa—Alluvial plain
ox—Oxbow lake
de—Delta

lg—Lagoon
be—Beach
ob—Offshore bar
nl—Natural levee
tf—Tidal flats
fp—Flood plain

Fig. 11.4

Hydrologically-significant distributions mapped from a radar mosaic of the Daren region of eastern Panama, and northwest Columbia: (b) The surface configuration. (Source: Viksne *et al.*, 1969)

attitude of bedding. Thus the drainage pattern overlay was the first to be developed from the imagery, being basic to most of the others. Clearly SLR has a part to play in regional hydrological studies especially where cloud cover restricts the use of aircraft as platforms for conventional photography.

At the scale at which the more remote systems operate most efficiently the complexity of the total hydrological problem, and the detail required by studies whether operational or research, comprise formidable hurdles to be overcome if remote sensing is to replace more direct 'contact' methods of assessing key variables. This is one of the

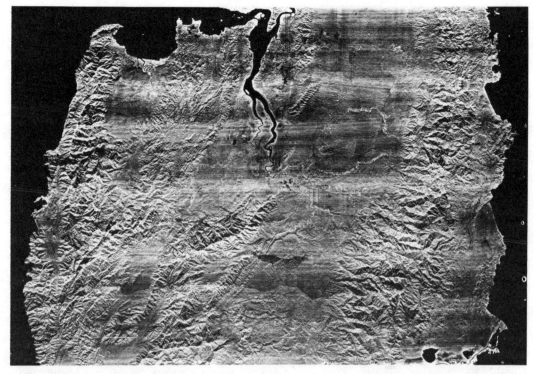

Plate 11.3
K-band radar mosaic of Darien (Panama) and north-west Colombia. (From Viksne, 1969)

relatively few fields for which further technological development is required before a useful potential might be realized. Remote sensing for hydrological purposes has been likened to the computer: here we have very powerful new tools whose potentials to yield large increases in efficiency in specifying and tackling environmental problems have yet to be fulfilled.

11.3.2 Satellite studies

It may seem surprising at first sight that studies of the hydrology of land surfaces by remote sensing from satellite altitudes are more immediately useful than those from surface or low altitude platforms. This is less surprising when we remember that, in general, as the scale is increased in hydrological studies, so the physical factors which are involved become less complex. Fortunately, large-scale, low resolution data have considerable scientific and economic value especially for planning the development of the water resources of the world.

The first Earth Resources Technology Satellite proved particularly useful as a test-bed for experiments with fairly direct hydrological applications. Indeed, the ERTS-1 Water Resources Working Group was in general agreement that the satellite's MSS observations provided much more useful and readily-applicable hydrological information than had been expected prior to its launch. However, before we summarize some of the ERTS-1 findings and Landsat inspired suggestions for the future, some notes are in order with reference to earlier satellite systems.

In the 1960's manned spacecraft of the relatively low-altitude Gemini and Apollo series secured some data of interest to hydrologists. This was mostly in the form of colour film images from hand-held cameras. There were, however, some additional multispectral photographs from a mounted array of four cameras on the Apollo 9 mission. Amongst other things such data showed that rainfall patterns may stimulate noticeable soil moisture differences, related to the distribution and amounts of rain which have fallen. These soil moisture differences lead to differences in soil surface reflectivity, so that relatively moist and dry areas may be identified through differences in photographic tone. Unfortunately, such information is difficult to evaluate in quantitative terms, and, in any case, can be obtained only if and when the skies have cleared after the rain. Active microwave systems able to penetrate cloud (like the SLR discussed in the previous section) have been proposed for future satellite operations, but none have been flown at satellite altitudes by the time of writing.

Weather satellites (e.g. Essa, Nimbus and Noaa) have provided interesting data with a hydrological content in thermal infrared wavebands and the microwave region using passive sensor systems. For example, drainage networks have been traced from high resolution infrared data in atmospheric window wavebands under cloud-free conditions at night, when water and the damp soils flanking water courses maintain a higher radiation temperature than the drier interfluves. Quantitative estimates can be placed upon the differences, but by and large the resolution (generally a few kilometres) has been too low for the results to be of much practical use except, perhaps in a few particularly inaccessible regions. The new Nimbus 5 Microwave Spectrometer (NEMS) gives more hope for the future. Reflectance patterns may be mapped by such means for soil moisture, snow cover, and ice type studies.

Returning to ERTS-based enquiries we may summarize the early accomplishments under six headings:

(a) Mapping drainage patterns and surface geology. It has been shown that stream channel development and network, stream length, and the location of ponds and lakes can be mapped from ERTS-1 data at a scale better than 1 : 250 000. The detection of geologic fracture zones, and especially the intersections of fractures, has potential for the future development of groundwater resources since it has been shown that high yield groundwater wells can be drilled along these fracture zones.

(b) Flood plain and inundation mapping. The mapping of inundated areas was one of the most striking hydrological applications of data from ERTS-1. Inundated areas mapped from ERTS-1 data corresponded closely to low altitude aerial mapping carried out for a comparison. A false-colour composite combining the blue and infrared bands was found to be the best discrimination. The effects of floods could be mapped even 7–12 days after the event. Some observations were made of the drying out of flooded areas. These and other studies should have a big impact on future mapping programmes of flood plains for zoning purposes.

(c) Assessments of surface water area. Surface water is prominent in the 0.8–1.1 μm observations. Water bodies as small as one hectare in size can be located, and rivers only 70 m across have been traced. The appearance or disappearance of land features around lakes or reservoirs can be used to evaluate water levels, and the volume of water stored.

(d) Wetlands mapping. Many important details of wetland zonation can be identified on Landsat imagery, for example in coastal wetlands where the following have been identified and mapped reliably at 1: 125 000: the marsh-water interface, upper wetland boundary, and plant communities consisting of different species of Spartina. Already Landsat data have been used in New Jersey to assist the state in its management of coastal resources.

(e) Assessments of water quality. Reflectance variation in river, lake, and ocean water bodies are due largely to variations in depth, suspended sediments, concentrations of pollutants, and/or biological activity of various kinds. We will return to this theme in greater detail in Section 11.4.

(f) Soil moisture studies. Areas of recent rainfall can be identified in semi-arid and arid regions, as from Gemini and Apollo photography . In areas of irrigation, large differences of soil moisture can be delineated. In particular, it is possible to compile an accurate synoptic inventory of fields under irrigation, or recently irrigated. Anomalous vegetation responses can be used to locate large and significant leaks from reservoirs and canals.

Other accomplishments have been achieved in ice and snow studies. We shall summarize them in the final section of this chapter. Certainly many of the Landsat based findings could be bettered in terms of resolution using airborne, not satellite-borne, sensor systems. The importance of the Landsat approach lies in the regularity and frequency of the satellite overflights, and the relative cheapness of each individual pass by comparison with the high costs of an average aircraft operation. There is little doubt that the use of Earth resource satellite techniques for studies in land surface hydrology will become very important if and when a fully operational satellite system becomes a reality.

11.4 Characteristics of water surfaces

11.4.1 Sea state

The term 'sea state' has been used widely for describing the characteristics of the wind-modified surface of the ocean. Measuring or profiling waves on large water bodies is important to a number of problems in oceanography. Radar research has revealed that a marked variation in microwave temperature brightness results from variations in ocean surface conditions. The technique works effectively until cresting waves form or spray becomes widespread. Some indications of sea state can be gained also from weather satellite visible imagery through areas of sun glint which may appear when clouds are absent and the alignment of sun and satellite permits. These indications are generally too coarse to be of much practical significance.

11.4.2 Water temperature and salinity

Knowledge of the sea-surface temperature distribution over large areas is important in the understanding of certain oceanic and atmospheric processes, for example certain details of the heat balance of the Earth/atmosphere system (see p. 182). Other useful studies have considered the detection and monitoring of ocean currents, upwelling zones and other thermal or motion systems. Such information may be important for naval operations, sea fishing, and commercial shipping activities, on account of known relationships between the thermal field of the ocean, and ocean circulation features, sea state, and the weather.

Apart from very localized studies of water temperature based on aircraft observations, some interesting work has been based on data from weather satellites, and more recently. Landsat and Skylab. Since only sporadic measurements are possible from shipboard, whilst aircraft are restricted by their operational range, a satellite system of water temperature mapping is highly desirable where broad-scale studies are contemplated.

Any unique interpretation of infrared radiance measurements from a satellite in terms of sea surface temperature requires a homogeneous target of known emissivity filling the radiometer field of view. Studies of sea surface temperatures using Nimbus satellite infrared data have employed synchronised cloud photographs to reveal the extent to which radiation temperatures of a target area are contaminated by the presence of clouds. These, of course, reduce the temperature levels. Aircraft-borne support systems have been used to assess the accuracy with which prominent features (e.g. the boundary of the Gulf Stream) are indicated by Nimbus isotherm maps. Fig. 11.5 shows a single scan line from a Nimbus HRIR analog record, along with an enlarged section of the analog record after digitization and the application of a numerical filter to remove oscillatory noise.

Since temperature patterns have been mapped not only on individual days in different parts of the world but also on a repetitive basis in selected areas in the search for secular patterns of temperature change, Fig. 11.6 illustrates an application of Nimbus HRIR data to the study of short-period temperature variations. Such studies reveal the dominant circulation patterns of water at different temperatures, and assist us in assessing the parts they play in related systems, for example the Earth/atmosphere energy budget. In the case of the Persian Gulf the incoming water from the Gulf of Oman compensates the strong local excess of evaporation over precipitation. The inflow

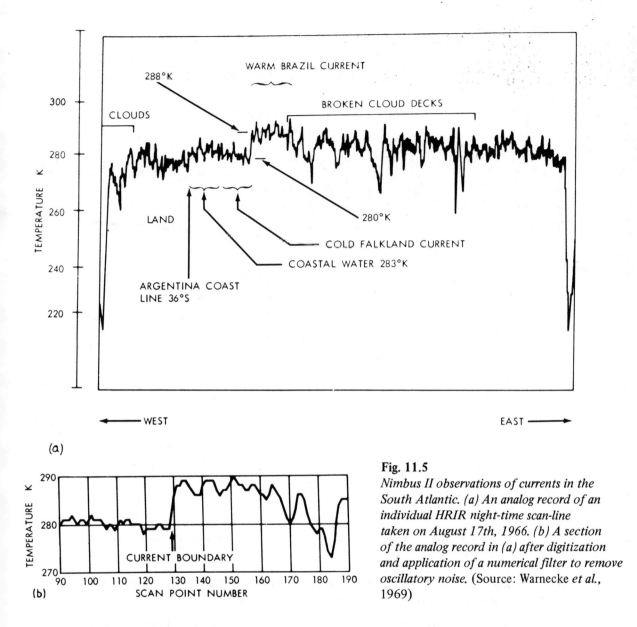

(a)

(b)

Fig. 11.5

Nimbus II observations of currents in the South Atlantic. (a) An analog record of an individual HRIR night-time scan-line taken on August 17th, 1966. (b) A section of the analog record in (a) after digitization and application of a numerical filter to remove oscillatory noise. (Source: Warnecke et al., 1969)

current spreads along the Iranian coast until it turns cyclonically along the Trucial coast. The water is warmed in this shallow region, and flows back eastward as warmer water to mingle with the currents along the Iranian coast.

More recently, staff at NESS have been working to develop methods for processing by computer data from operational weather satellites for sea-surface mapping. The object is to map temperature distributions by more objective means. Radiation tem-

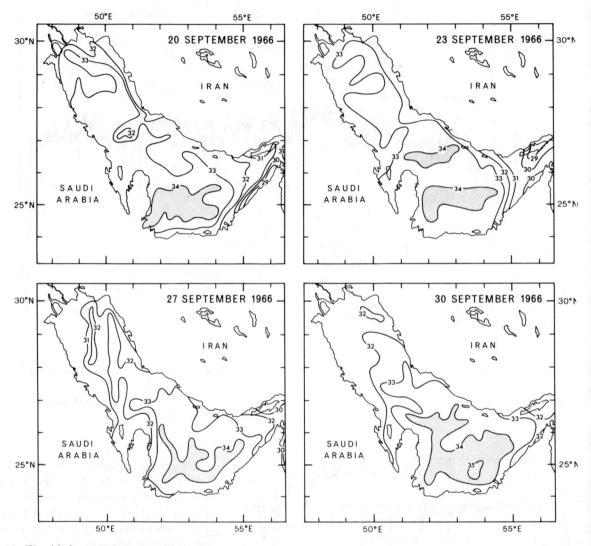

Fig. 11.6

Nimbus II HRIR observations of surface temperatures across the Persian Gulf revealing a cycle of warming (a-d) associated with the strengthening water circulations (e). (Source: Szekielda *et al.*, 1970)

perature frequency distributions for selected latitude/longitude squares are used to assess the presence or absence of cloud. Where cloud cover is complete, or totally absent, the histograms are unimodal. In partly cloudy regions they tend to be bimodal or multimodal. The highest temperature mode corresponds to surface temperature, the lowest to extensive cloud contamination. A daily temperature mapping programme is planned for the future, reserving only those temperatures which relate to the surface itself.

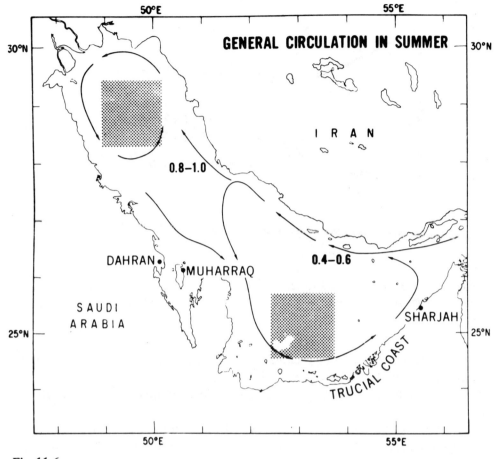

GENERAL CIRCULATION IN SUMMER

0.8–1.0

I R A N

0.4–0.6

DAHRAN

MUHARRAQ

SAUDI
ARABIA

SHARJAH

TRUCIAL COAST

Fig. 11.6e

11.4.3 Pollution and salinity

Natural variations in the salt content of the oceans have been investigated by the Agricultural and Maritime University of Texas. Workers there have computed the emissivity and absorptivity of water in the microwave region as functions of wavelength, water temperature, dissolved gases, and salinity. Fig. 11.7 shows the variations they discovered in apparent temperature as functions of water temperature for fresh and salt water. These could be useful for studies of estuarine water circulations.

Contamination of water bodies by human activities lends itself more readily to investigation by remote sensing techniques since the effluents of commercial and industrial activity often contrast sharply with the waters which receive them. Aircraft have been used in many instances to identify and map the discharge patterns from sewage outfalls and industrial complexes. Oil spillages from ships at sea, and oil

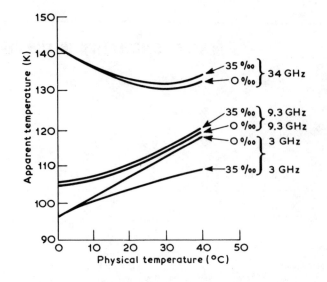

Fig. 11.7
Theoretically predicted curves of sea surface brightness temperature (vertical axis) versus physical temperature (horizontal axis) for different salinities (⁰/oo) at selected wavelength of emitted microwave radiation.

leaks from maritime wells can be located with particular ease from aircraft — and, potentially, from satellites. There are four reasons why this is so:

(a) The emissivity for petroleum products is significantly higher than for a calm sea surface.
(b) Crude oil pollutants have increasing emissivities with increasing gravity.
(c) The radiometric response of oil varies little with time of day or the age of the pollutant.
(d) It is possible to design reliable microwave systems to map oil pollution.

11.5 Frozen surfaces

Two aspects of ice and snow distributions transcend the rest, namely the significance of frozen water within the hydrological cycle, and the influence of sea ice on maritime activities. A substantial portion of the world's fresh water resource is stored temporarily in the form of snow especially in high mountain regions. Knowledge of the distribution of snow fields and their volumes in terms of water equivalents is required so that we can improve forecasts of stream flow and water storage. These in turn strongly affect power generation programmes, irrigation, water quality control, river management, manufacturing, recreation, and many other factors which contribute to a nation's economy. Sea ice imposes a seasonal control on shipping in many regions of the north-

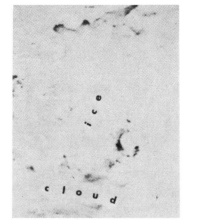
26 April 1970

25 May 1970

24 June 1970

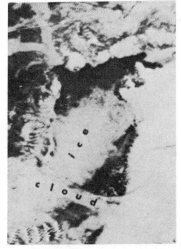
9 July 1970

31 July 1970

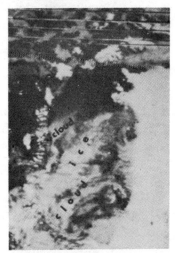
9 August 1970

2 September 1970

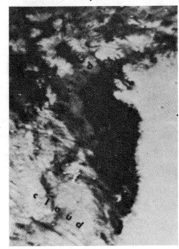
8 September 1970

9 September 1970

Plate 11.4

Nimbus IV Image Dissector Camera System (IDCS) pictures of Baffin Bay. These are examples of the daily data supplied from weather satellite to the U.S. Navy. They are used to provide ice forecast support, and to reduce aerial reconnaissance requirement. (Courtesy, NASA).

ern hemisphere. The opening and closing of ports is influenced thereby, and certain navigation routes can be used for only a part of each year, with or without the use of icebreakers.

Scandinavian countries are amongst those which have developed airborne techniques for estimating the volumes of snow accumulated on watersheds which serve hydro-electric stations at lower altitudes. The North American nations, Canada and the U.S.A., have long operated an expensive programme of ice reconnaissance in the St. Lawrence region and northern coastal waters using aircraft to spot natural leads and passages through the ice. Both types of programmes could be superseded eventually by satellite-based methods (Plate 11.4).

Already quite detailed snow boundary mapping is possible using data from the Noaa VHRR infrared sensor. This system, which gives a resolution marginally less than 1 km, will make it possible for much more detailed displays than those in Plate 11.4 to be compiled in the future. Other studies of the albedoes of snow fields have shown that the age of snow can be deduced from the changes in reflectivity which take place after it has fallen.

Studies of sea ice from weather satellites have been more restricted owing to the poor illumination of high latitudes in the local winter season and the consequent almost total dependence upon thermal emission patterns. The resolution of weather satellite data has generally precluded the obtaining of information on the age and conditions of sea ice, but it has been possible to use such data to determine the extent of the ice, and, within a little, the stage of its formation or breakup. An early Tiros-based programme ('TIREC') revealed that even quite coarse-resolution data could lead to significant economies in standard ice reconnaissance operations. Satellite data were analysed for areas which might be surveyed usefully in more detail at aircraft altitudes. Since then weather satellite data have played their part each year in North American ice reconnaissance.

We may conclude this chapter with reference once again to ERTS-1 as a harbinger of future developments. Under optimum (cloud free) conditions, elevations of snow lines have been estimated to the nearest 60 m from ERTS evidence; in some regions (e.g. the Northern Cascades in the State of Washington) the areas covered by snow been estimated to a proven accuracy of 96 per cent. In other areas more detail was obtained from ERTS-1 snowline mapping than is obtained commonly from operational aircraft surveys, in approximately the same time for analysis. Multispectral data for glaciers have permitted distinctions to be drawn between surging glaciers and non-surging glaciers through the associated morainal forms. Melt rates and run-off volumes can be assessed even in quite inaccessible regions by sediment plumes extending away from glaciers. Worked examples include glaciers in southern Alaska, melting into the Alaskan Gulf.

As in so many other fields where remote sensing techniques have clear potential applications, a very significant task for the future lies in educating new users to their benefits. Perhaps the most urgent need of all in hydrology and its germane sciences is for a clearer definition of the spatial, temporal and spectral frequencies of the data required for the achievement of realistic aims. Specific problems, for example in water resource evaluation and management, cannot be solved though the remote sensing approach until the potential of remote sensing has been realized and suitable problems have been identified.

References

Allison, L.J. *et al.* (1972), 'Air-sea interaction in the tropical Pacific Ocean,' NASA Technical Note, TN D-6684, NASA, Washington.

Chorley, R.J. (ed.) (1969), *Water, Earth and Man,* Methuen, London.

Freden, S.C. and Mercanti, P. (1973), Discipline Summary Reports, Vol. III, *Symposium on significant results from ERTS-1,* Report No. X-650-73-155, NASA, Greenbelt, Maryland.

Grinstead, J. (1974), 'The measurement of areal rainfall by the use of radar', in *Environmental Remote Sensing: applications and achievements,* Barrett, E.C. and Curtis, L.F. (eds.), Edward Arnold, London, pp. 267-284.

Martin, D.W. and Scherer, W.D. (1973), 'Review of satellite rainfall estimation methods', *Bull. Amer. Met. Soc.,* **54**, 661.

McClain, E.P. (1970), 'Applications of environmental satellite data to oceanography and hydrology', ESSA Technical Memorandum, NESCTM 19, U.S. Department of Commerce, Washington.

Nagle, R.E. and Serebreny, S.M. (1962), 'Radar precipitation echo and satellite cloud observations of a maritime cyclone', *J. Appl. Met.,* **1**, 279.

Painter, R.B. (1974), 'Some present uses of remote sensing in monitoring hydrological variables', in *Environmental Remote Sensing: applications and achievements,* Barrett, E.C. and Curtis, L.F. (eds.), Edward Arnold, London, pp. 285-298.

Plessey Radar (1971), 'The Dee weather radar project', Publication No. 6776, The Plessey Company Limited, Weybridge.

Szekielda, K.H., Allison, L.F. and Salomonson, V. (1970), 'Seasonal sea surface temperature variations in the Persian Gulf', Goddard Space Flight Center, Report No. X-651-70-416, NASA, Greenbelt, Maryland.

Viksne, A., Liston, T.C. and Sapp, C.D. (1969), 'SLR reconnaissance of Panama', *Geophysics,* **34**, 54.

Warnecke, G.L., McMillin, M. and Allison, L.J. (1969), 'Ocean current and sea surface temperature observations from meteorological satellites', NASA Technical Note, D-5142, NASA, Washington.

 **Soils and landforms**

12.1 Introduction

Soil is of major importance to mankind because it supports the plants and animals that provide man with food, shelter and clothing. It is also of importance as the foundation material for the buildings in which we live. Different concepts of soil have therefore been developed by agriculturalists and engineers.

The agriculturalist finds the whole of the soil profile significant to plant growth: the surface (A) horizon because it is the seat of biological activity and the principal source of nutrients; the subsoil (B) horizon because it affects drainage, soil moisture retention, aeration and root development of plants. Both A and B horizons are important in respect of their influence on soil temperatures. The underlying parent material (C horizon) is mainly significant because it may contain weatherable minerals yielding nutrients and its texture may affect permeability. Deeper parts of the regolith (D horizons) are generally of less importance to plant growth.

In contrast with the agricultural concepts of soil, the engineer normally defines soil as the unconsolidated sediments and deposits of solid particles derived from the disintegration of rock. Therefore in civil engineering the term soil includes all regolith material and a sharp demarcation between rock and soil is no longer made. In fact, the engineer is normally more concerned with what the agriculturalist would term the C and D horizons. From the engineering standpoint soil can be regarded as a three-phase system which is in dynamic equilibrium. The three phases are: soil (particulate organic and inorganic material), liquid (consisting of soil solutions containing various salts) and a gas phase containing soil air with changing amounts of oxygen, nitrogen and carbon dioxide. The equilibrium between these phases changes continuously and the engineer studies the interactions of the phases and the physical properties of soils affecting soil engineering e.g. soil stability.

12.2 The mapping of soils

The recognition of soils as organized natural bodies is a development of nineteenth-century science and is generally credited to the Russian scientist Dokuchaiev who first explained the relationships between soil profiles and the environment. It was not until the 1920s, however, that soil units were eventually mapped on the basis of the

soil profile using A,B,C horizon notations. With further experience of soil mapping
it was recognized that soils were very variable in their properties and several kinds of
soil might occur within a mapping unit. Even though a major part of a mapped unit
might accord with the definition of a particular soil type, not all of the area would
conform. Thus it became usual to recognize that a percentage (up to 15 per cent) of
soils within a mapping unit would show variation from the described attributes of the
unit. In general, the user wishes within-class variance to be as small as possible, or
below a certain threshold value. A number of studies of soil variability have been made
including variances and intra-class correlations for various topsoil properties in Oxford,
England which showed mechanical properties were acceptable in terms of within-class
variance, whereas chemical properties were not.

Studies of soil units by remote sensing methods must, therefore, take into account
both different concepts of soil bodies held by agriculturalists and engineers and the
variability of topsoil characteristics, particularly in respect of chemical properties.

The soil categories used by agriculturalists that are significant to remote sensing
studies include those of great soil groups, sub-groups, soil series and soil phases. The
great soil group brings together soils having similar horizons arranged in similar sequence
e.g. Typorthod (Podzol). Thus a broad zone of Podzolic soils extends across northern
Canada and Russia over a variety of rock types. This unit is, therefore, suitable for
small scale (1: 1 000 000 or less) or regional maps. At medium scales (c. 1: 50 000–
1 100 000) the soil series category is more commonly used. A soil series consists of a
group of soils of similar profiles derived from a particular parent material. Thus different
soil series are found on different types of underlying rock and sediment. The soil series
is often subdivided into soil phases. The basis of subdivision into phases may be any
characteristic or combination of characteristics potentially significant to man's use or
management of soils. The most common differentiating attributes for soil phases are
slope, degree of erosion, depth of soil, stoniness, salinity, physiographic position and
contrasting layers in the substratum.

Soil classifications for engineering purposes are primarily based on the sieve size
of the soil material. Thus the extended Casagrande Soil Classification as used by
engineers has as its major divisions, coarse-grained soils and fine-grained soils, subdivided
into gravelly, sandy and fine-grained soils. Finer subdivisions in the classification use
criteria such as uniformity of deposit, grading of deposit, plasticity, shrinkage or
swelling properties, drainage characteristics and dry density.

12.3 Remote sensing studies by photographic systems

It has been shown above that engineers and agriculturalists adopt different definitions
and classifications of soils. As a result remote sensing techniques are required to serve
several purposes based on different definitions of soil and different basic concepts. The

first major post war study of the recognition of soil conditions made at Purdue University, U.S.A., was mainly aimed at identification of parent materials and was chiefly of value to those interested in soils engineering. It was based on two assumptions. First, that once photographic characteristics for soils at one site had been determined the identification of similar sites elsewhere on the photographs could be used to identify similar soils. Second, that by study of the aerial photograph such features as landform, surface colour (or tone), erosion, slope, vegetation, land use, micro-relief and drainage patterns could be used to deduce the general character of the soil. Further development of this latter approach by the International Training Centre for Aerial Survey, Delft, Holland led to a technique of pedological analysis. In this, each element of the landscape was mapped out separately (e.g. slope, vegetation, land use) and a map was prepared from a composite of the boundaries. The boundaries which were coincident with more than one element of the landscape were given additional weight in the final interpretation of soil boundaries.

The widespread recognition of the close association of soils and landforms promoted interest in landscape mapping as a means of reconnaissance soil mapping. Research was based on the idea that knowing detailed characteristics for one terrain unit, one can apply them to another analogous terrain unit elsewhere. The problem facing terrain analysis was (and continues to be) one of classification and definition of the units to be recognized on the photograph. It was necessary to have some means of identifying the terrain unit and storing data relevant to the unit so defined so that other workers could recognize it elsewhere. Thus a system of hierarchical classification was proposed as shown in Table 12.1.

The land system approach has been widely applied, for example, in Australia by Land Research and Regional Survey teams, South Africa by engineers and Africa by agriculturalists. It is best applied to those areas which display distinctive and well-

Table 12.1
Terrain unit classification

Unit	Description
Land element	The simplest part of the landscape; uniform in lithology, form, soil and vegetation.
Land facet	Consists of one or more land elements grouped for practical purposes
Land system	Recurrent pattern of genetically linked land facets
Land region	Consists of one or more occurrences of land system local forms which are generally contiguous
Land division	An assemblage of surface forms expressive of a major continental structure (morphotectonic)
Land zone	The world extent of a major climatic type

defined facets of landscape. Where units are ill-defined and where cultivation has largely destroyed the natural vegetation cover it is less easily applied.

The development of computing techniques has led to investigations of the parametric description of land forms on an entirely numerical footing. Computer analysis of land form now seems to be within our grasp and future developments of the land system approach are likely to be in the form of automated cartography using some form of supervised classification based on knowledge of soil-landscape relationships.

The accuracy of soil mapping from the air photograph has proved to be very variable. Comparisons of soil maps made from ground observations with photo-interpretation maps have revealed that increasing complexity of geological deposits or increasing importance of properties not closely associated with landforms greatly reduces the accuracy. Also there is increasing awareness that the timing of air photography is critical. Photographs taken at inappropriate times may provide an unsatisfactory yield of information. Under European conditions photographs to record soil-tone patterns and soil changes should be taken in winter and spring, the best months being March and April (Plate 12.1). Crop patterns which reflect soil boundaries are best recorded in July and early August. These requirements must,

Plate 12.1 (a)
Soils on a former estuarine marsh and adjacent upland, near Martin, Lincolnshire: (a) under crops, July 1971.

Plate 12.1 (b)
Bare soil, April 1971. (Note the clear portrayal of the patterns of creeks which formerly traversed the marsh. The tonal variations in the photograph along the creek lines are primarily due to variations in soil texture and moisture content. (Source: Cambridge University Collection: Copyright reserved).

however, be fitted in to the period when conditions are suitable for aerial survey and so the planning of flights and placing of contracts should be carried out well in advance.

Suitable statistical methods for the testing of the goodness of soil boundaries drawn by photo interpretation have become necessary following the increasing use of air photography. Multiple correlated attributes of the soil profile can be reduced to a single variate expressing a large proportion of available information by principal component analysis. This variate, the first principal component, can be plotted against distance on a linear transect and approximate positions of maximum (positive) and

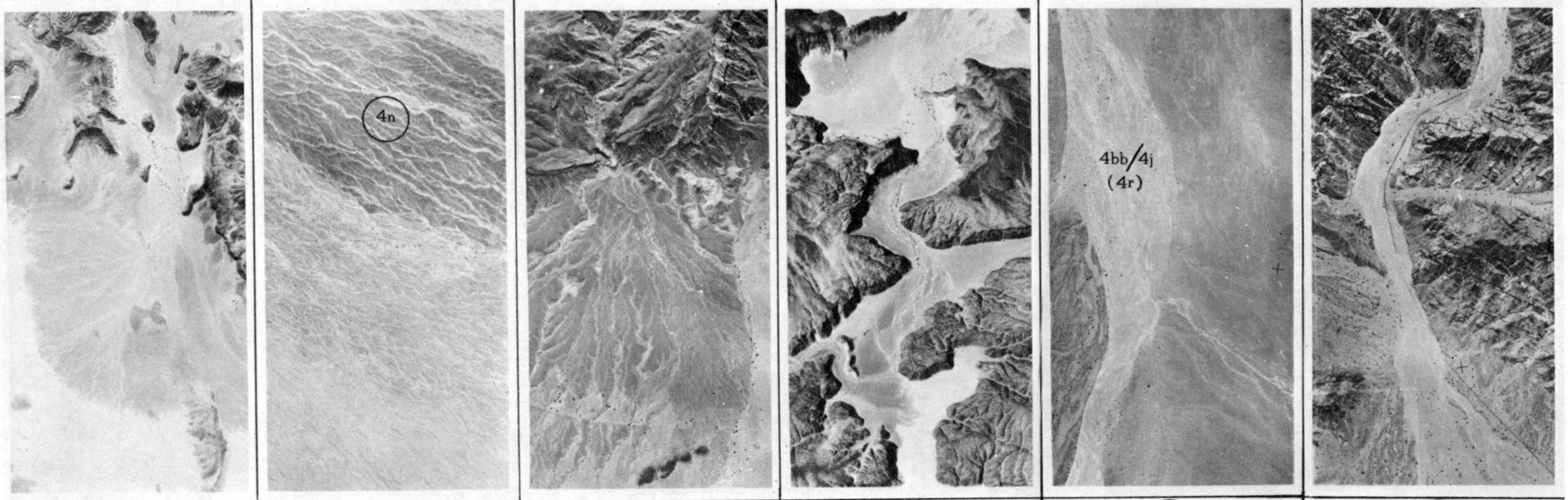

4n	4n	4o	4q	4r	4s
Gradual gravel fans of granite or sandstone, liable to occasional floods but usually free-draining.	Gradual gravel fans of granite or sandstone; liable to occasional floods but usually free draining. Wadi Araba region.	Gravel covered plains usually derived from limestone. Drainage free. May be alkaline.	Shallow sandy wadis.	Wadis of coarse sandy alluvium.	Bouldery wadi bottoms among granite hills, subject to sudden and violent floods.
Very sparse brush.	Very sparse brush.	Sparse scrub of Haloxylon salicornicum, Traganum nudatum and Suaeda vermiculata.	Brush of Artemisia judaica.	Brush of Haloxylon persicum, Retama, etc.	Sparse brush.
Marked fan formation in medium grey tone	Fan formation in medium to dark grey tone. Distinguishable from 4k by a rather coarser drainage pattern.	Medium grey tone with round moulded appearance.	Light grey tone with drainage channel showing as string of dark spots.	Light grey overall tone with drainage channel showing in lighter smooth appearance.	Light grey tone of fine texture. Scattered vegetation showing as dark spots.

Plate 12.2

Soil range conditions in Jordan. (Courtesy Aerofilms, Hunting Surveys Ltd., Boreham Wood, Herts.)

minimum (negative) slope found by inspection. These positions are then pin-pointed using Student's *t* statistic. They represent the points where the soil boundaries, the maximum rates of change of soil with respect to distance, cross the transect. Beckett (1974) has recently reviewed the statistical assessment of resource surveys by remote sensors, particularly in respect of the precision of information and the costs of obtaining data on soil conditions at different scales.

The use of colour photography as an aid in soil mapping only became important after 1960 following advances in technology by film manufacturers. Studies of the use of colour film for engineering soil and landslide studies showed that it gave a saving in interpretation time of fifty per cent in the U.S.A. Research by British and Russian scientists indicates that colour aerial photography yields more data on soil conditions than panchromatic photography. Further work has shown, however, that soil colours as recorded in the field and on aerial photography do not correspond exactly in their Munsell colours, indeed overlap of hue, value and chroma is quite frequent. The influence of different soil hues on the optical densities of aerial collour and aerial infrared colour has subsequently been studied by densitometric methods. Statistical tests applied to density data obtained from a sample of twelve soils have shown that the soils can be separated into two groups on the basis of significant differences in densities. One group consisted of those soils with low chroma (soils grey or neutral in colour) which were best distinguished by infrared colour. The second group included soils with high chroma (intense soil colour) which were best distinguished with colour film.

Infrared colour transparencies require suitable techniques for viewing and annotation and suitable light tables and filter frames have been developed as more and more infrared colour has been used. Techniques have also been developed for the production of panchromatic internegatives and positive prints from infrared colour film. There remains, however, the fact that transparencies are less convenient for use in manual photo-interpretation and this may partly explain the somewhat slow adoption of infrared in user agencies despite its considerable information yield in respect of soil conditions and crop vigour. A summary of the advantages and disadvantages of different types of film used in aerial surveys is given in Table 4.2.

In order to communicate the photographic characteristics of soil and terrain units to other users photographic keys of various kinds have been devised. Many are available for use, including keys for vegetation, landforms, forest sites, and soil conditions. An example of a photographic key for soil and range conditions in Jordan is shown in Plate 12.2. The continuing interest in site selection is reflected by recent publications dealing with terrain evaluation, some of which are listed in the references given in this chapter.

12.4 Spectral signatures of soils

Since soils are composed of mineral and organic constituents together with varying amounts of soil moisture it can be assumed that different soils will show different reflectance characteristics as these constituents vary in amount and kind. There was, therefore, early work by soil scientists on the reflectance characteristics of different soil types, particularly in the infrared part of the spectrum. These studies made in the 1950's were not related to remote sensing techniques but with the increasing use of multipsectral

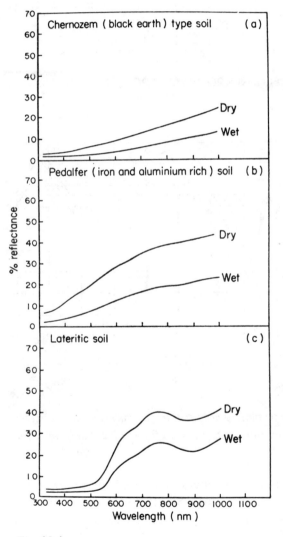

Fig. 12.1
Spectral characteristics of soils. (Source: Condit, 1970)

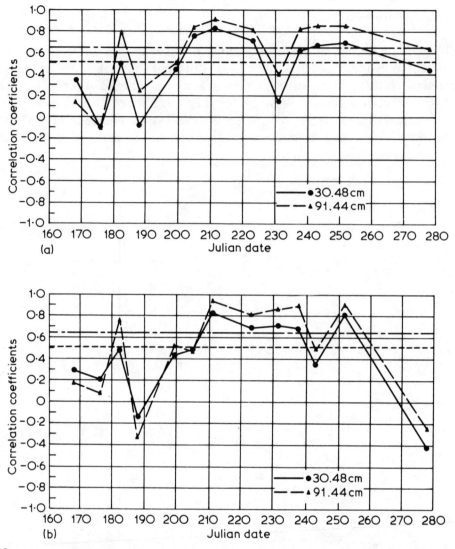

Fig. 12.2

Mission correlation coefficients plotted against time for the available soil moisture in the sorghum versus film density (a) colour infrared film, and (b) green filtered black and white film. (Source: Werner *et al.,* 1973

aerial cameras and multispectral scanners there has been renewed interest in spectral signatures of surface objects, including soils. Recent laboratory measurements within the wavelengths of 0.32–1.0 μm have been made on soils in wet (almost saturated) and dry (oven dried at 43°C) states.

From some 160 sets of curves three general shapes emerged (Fig. 12.1). Type 1 curve was charateristic of chernozem-like (Mollisol) soils, Type 2 of pedalfer soils (Spodosol), Type 3 of lateritic soils (Oxisol). An important part of the analysis of these results showed that reflectance measurements made at only five wavelengths would suffice to predict with sufficient accuracy the other 30 wavelengths measured. It was recommended that the five selected wavelengths should be 0.4, 0.5, 0.64, 0.74 and 0.92 μm. Much more extensive studies of spectral signatures have now been made in the field as well as in the laboratory under NASA contracts. These consist of airborne and ground measurements over variable parts of the spectrum within the range 0.3–45 μm. Similar studies have been carried out by Russian scientists. For example, densitometric studies of spectral reflectance of sands in the Karakum Desert have shown that isopleths drawn by trend surface analysis of the spectral data can be used to distinguish three types of sand with differing minerological characteristics.

Spectral reflectance data has also been examined for the purpose of evaluating soil moisture and water use by plants by airborne remote sensing methods. Relationships were investigated between remote sensing imagery and soil moisture in grain sorghum. Three spectral ranges were used, 0.47–0.61 μm (Green), 0.59–0.70 μm (Red) and 0.68–0.90 μm (near Infrared). It was found that soil moisture differences could best be detected in the red portion of the spectrum. The near infrared region of the spectrum gave little information concerning the soil moisture status. The green spectral band gave good results during the time of active plant growth. Examples of correlation coefficients plotted against time for the available soil moisture in the sorghum versus film density for red filtered (8403 film) and infrared (2424 film) are shown in Fig. 12.2.

12.5 Remote sensing of soils by non-photographic systems

12.5.1 Scanning in visible and near infrared wavelengths

The multi-spectral scanner (MSS) used in ERTS-1 provided data in the visible and near-infrared wavelengths with pass bands of (4) 0.5–0.6, (5) 0.6–0.7, (6) 0.7–0.8, and (7) 0.8–1.1 μm. Preliminary assessments of the results obtained from this imagery (NASA., 1973) include studies of soil conditions. The ERTS-1 investigators have found that interpretation of satellite imagery permitted identification of subirrigated areas, together with complexes of 'choppy sands' and 'sands range' sites in the Sand Hills region of Nebraska. This region occupies the north-central one third of the state and is centrally located in the Great Plains. The region is composed of approximately 20 000 square miles of aeolian sand dunes interspersed with nearly level valleys. Colour composites derived from MSS bands 4,5 and 7 were used to distinguish different range sites and forage density. The sands range sites carried Typic Ustipsamment soils whereas the subirrigated areas were associated with Aquic Haplustolls and Typic Haplaquolls. It was noted that burned areas could be accurately measured from the imagery. Also areas of wind erosion could be identified.

ERTS-1 imagery of Obion County, West Tennessee has been studied with the aim of making a delineation of soil associations. It was found that major soil associations such as the Memphis occurring at the break between loess soils and the delta soils of the Mississippi floodplain can be identified. A high speed digital scanning micro-densitometer was used to scan the ERTS image of the area using a 25 μm raster. Subsequently the computer printouts were able to separate different soil associations. (Fig. 12.3). It was found that channel 7 was the most useful one for such studies. Se-

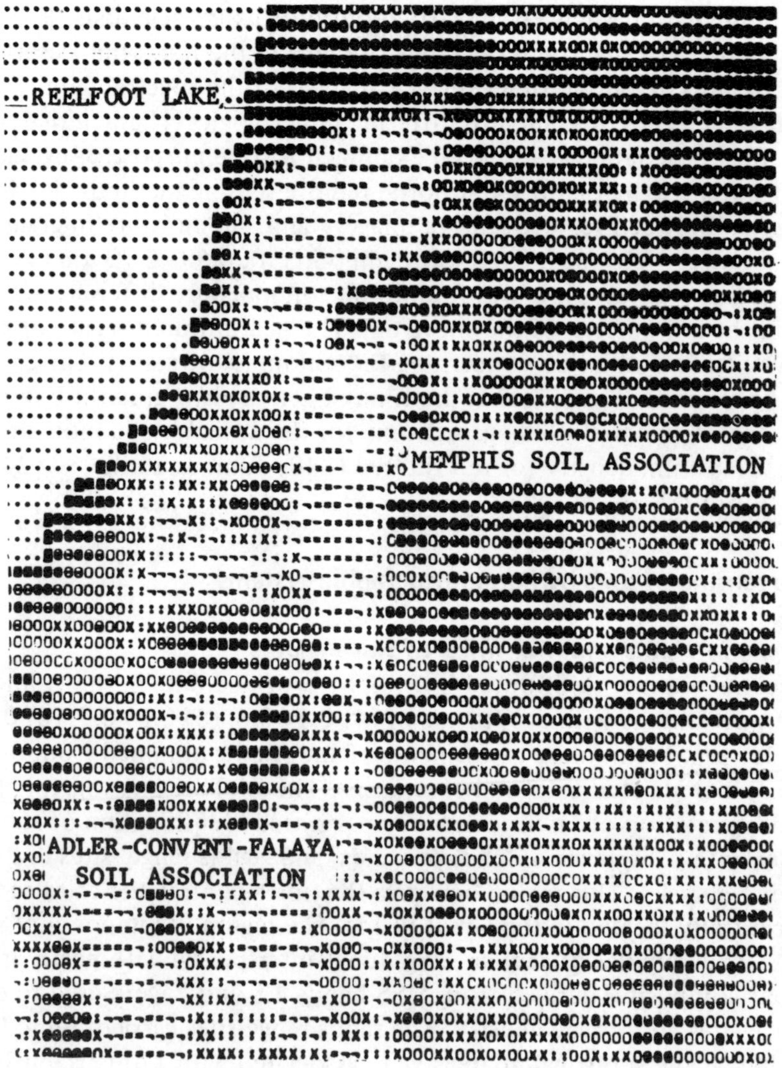

Fig. 12.3
ERTS image and computer print-out of soil association map. (Source: Parks and Bodenheimer, 1973)

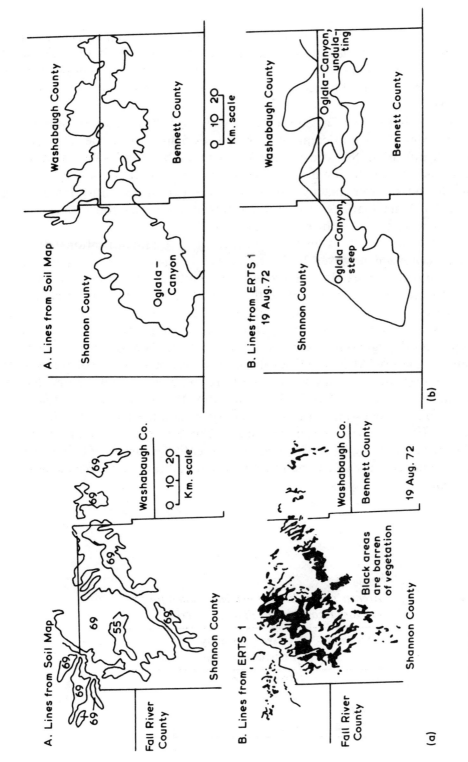

Fig. 12.4
Comparisons between existing soil maps and ERTS-1 imagery. (Source: Weston and Myers, 1973)

paration of small soil associations in intensive row crop agricultural areas was found to be much more difficult at the scale of Landsat imagery. In this study the reflectance was thought to be at a minimum at low moisture contents (about 2 bars tension) whereas maximum reflectance was obtained at moisture levels slightly below field capacity (1/3 bar tension).

Studies of Landsat imagery of South Dakota have also shown that soil associations can be identified from film colour composites of bands 4,5 and 7 when viewed over a light table with magnification. Specific comparisons between boundaries shown on the Landsat composites and soil association maps are shown in Fig. 12.4. It was found that soil association boundaries were mainly distinguished through the identification of boundaries around landscape features. Vegetation differences were also found to assist the placing of boundaries around the soil associations.

The work outlined above was mainly concerned with soil identification. Other studies have been made, however, to determine whether general terrain units such as water, exposed rock, grasslands, coniferous forest and lowland marsh areas can be identified. An example of such a study is that carried out in respect of terrain classification maps of Yellowstone National Park. Training sets of the five categories just described were defined by ground observations. Samples were selected for each of the terrain types and the signature of each type was extracted from the digital record of the Landsat imagery. At first individual (i.e. from one band only) signatures were obtained. Later signatures from several bands were combined. The optimum channels for classification were then determined by calculating the average probability of misclassification (i.e. an estimate of the percentage of points which will be incorrectly classified in a terrain recognition map). The results were as shown in Table 12.2.

It will be seen that using bands 5,6,7 was only slightly less accurate than using all four bands. Thus economy of computer effort would indicate that three bands only might be used.

Table 12.2
Average probabilities of misclassification for five category recognition (Yellowstone Park)

MSS Bands	Average probability of misclassification
5	0.086
5, 7	0.031
5, 7, 6	0.028
5, 7, 6, 4	0.027

12.5.2 The use of thermal infrared and microwave scanners

Although the scanners in the visible and near infrared wavelengths are producing good results there is considerable research into the applications of thermal infrared and microwave scanners for soil studies. The use of thermal infrared for soil temperature studies has been investigated by NASA investigators in aircraft studies.

Detailed soil surveys were related to the thermal infrared imagery in a manner which suggested that surface soil temperatures can be indicative of subsurface soil conditions. Such results are to be expected in view of the effects of soil texture, composition, porosity and moisture content on soil temperature regimes in different soils. Thermal infrared scanning is particularly sensitive to moisture content at the very surface of

Plate 12.3
Infrared linescan imagery of esturine lowland near Huntspill River north of Bridgwater. Note dark tones of water and clear representation of field drains by dark lines of lower temperature. (Courtesy, Royal Radar Establishment, Malvern.)

soils. This is particularly the case when surface winds evaporate moisture from the surface layers. In these circumstances the cooling of the surface in exposed areas contrasts with the higher surface temperature in areas protected from the wind. An example of these conditions is shown in Plate 12.3 which shows part of an area of Somerset under conditions of 11°C air temperatures and 4.5–7.9 m s^{-1} wind strength This is an area of reclaimed estuarine clay which is under pasture and requires extensive field drainage. The open drainage ditches are readily distinguished by dark tones relecting the lower temperatures of the water surfaces in relation to the land. The lines of buried field drains can also be detected as a result of the wetter soil conditions along these sub-surface channels. The white toned areas in the lee of field boundaries reflect the shelter effects of the boundaries and the lower evaporation and hence higher surface temperatures existing in these sheltered places.

In view of the effects of moisture on the dielectric characteristics of surface materials there is increasing interest being shown in microwave sensors for the detection of moisture states. Also the dynamic nature of water conditions in the landscape makes the microwave sensors attractive because they can be used on a repetitive basis due to their relative immunity from atmospheric effects.

NASA has investigated the capability of airborne microwave radiometers to monitor soil moisture. Regression analysis has indicated a $-2.15°$ C/ per cent moisture variation in apparent temperature for the 1.42 GHz radiometer with vertical polarization. A relationship of $-1.5°$ C/ per cent moisture was obtained for vegetated fields using 19.4 GHz with horizontal polarization. These relationships must, however, be treated with some reserve since it would appear that they are based on moisture per cent by dry weight of soil rather than moisture per cent on a volume basis. Such data is capable of producing spurious correlations because different soil materials have different specific gravity values. As a result the same amount of water in similar volumes of organic soil and mineral soil will appear as a much higher percentage in the organic soil when expressed on a dry weight basis.

The quality of the ground data must also be questioned in recent Russian studies of microwave radiation data in relation to soil moisture content. In this work data obtained by the Russian satellite Cosmos 243 was related to soil moisture content as measured at meteorological stations in various parts of the Soviet Union. It was found that correlations between moisture content and radiometric data were highest for wavelengths of 8.5 cm. However some of the soil moisture estimates were only made on the basis of the visual appearance of the soil.

Interest in microwave methods also springs from the fact that in dry land sediments there is a penetration of the surface layers. Theoretical calculations show that penetration of 1 m or greater may be achieved (Fig. 12.5). Thus there is the possibility that development of microwave sensors will provide methods of integrating not only surface characteristics but also subsoil characteristics of soils. Such sensors would find considerable application in arid and semi-arid areas of the world. In these regions underlying structures may be revealed which have economic significance in respect of oil exploration. The sub-surface conditions are also of importance in planning and construction of irrigation schemes.

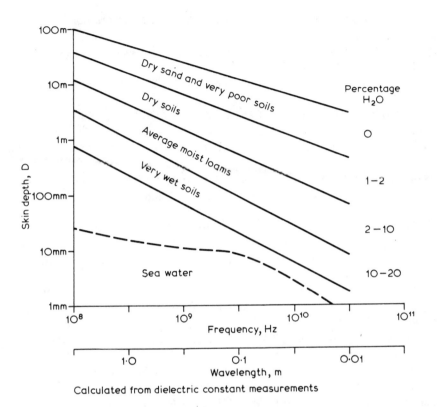

Fig. 12.5
Microwave penetration of soils.

References

Avery, G. and Richter, D. (1965), 'An airphoto index to physical and cultural features in the eastern United States', *Photogrammetric Engineering,* **31,** 896.

Basinski, J.J. (1959), 'The Russian approach to soil classification and its recent development', *J. Soil Sci.,* **10,** 14.

Beckett, P.H.T. (1974), 'The statistical assessment of resource surveys by remote sensors, in *Environmental Remote Sensing; applications and achievements,* Barrett, E.C. and Curtis, L.F. (eds). Edward Arnold, London pp. 11-29.

Beckett, P.H.T. and Webster, R. (1965), 'A classification system for terrain. Units and principles', Military Engineering Experimental Establishment, Christchurch, England. Report No. 872, p.29.

Belyakova, G.M. *et al.* (1971), 'Study of microwave radiation from the satellite Cosmos 243 over agricultural landscape', *Coklady Akademic Nauk,* U.S.S.R., **201**, 837.

Brink, A.B.A. (1963), 'Airphoto interpretation applied to soil engineering mapping in South Africa', *International Archives of Photogrammetry,* **14**, 498.

Casagrande, A. (1948), 'Classification and identification of soils', in *Transportation,* American Society of Civil Engineers.

Christian, C.S. (1958), 'The concept of land units and land systems', *Proceedings of the 9th Pacific Science Congress, Bangkok,* 1957, **20**, 74.

Condit, H.R. (1970), 'The spectral reflectance of American soils', *Photogrammetric Engineering,* **36**, 955.

Curtis, L.F. (1971), 'Soils of Exmoor Forest', Report Soil Survey Gt. Britain, Rothamsted Experimental Station, Harpenden, p. 77.

Curtis, L.F. and Trudgill, S. (1974), 'The measurement of soil moisture', *British Geomorphological Research Group Bulletin,* **13**.

Curtis, L.F. (1974), 'Remote Sensing for Environmental Planning Surveys', in *Environmental Remote Sensing; applications and achievements,* Barrett, E.C. and Curtis, L.F. (eds.), Edward Arnold, London, pp. 87-109.

Gaveman, A.V. and Liverovskii, Yu.A. (1953), 'Aerial photography in soil mapping', *Pochvovedenie,* **3**, 1.

Gerbermann, A.H., Gausman, H.W. and Wiegand, C.L. (1971), 'Color and color I.R. films for soil identification', *Photogrammetric Engineering,* **37**, 359.

Jenkins, D.S., Belcher, D.J., Gregg, L.E. and Woods, K.B. (1964), Technical Development Report 52, U.S. Dept., Commerce, Civil Aeronautics Administration, Washington.

Jumikis, A.R. (1962), *Soil Mechanics,* Van Nostrand, New York.

Kroll, C. (1973), *Remote Monitoring of Soil Moisture using Airborne Microwave Radiometers,* Texas University Remote Sensing Center.

Landis, G.H. (1955), 'Concept and validity of association photographic interpretation keys in regional analysis', *Photogrammetric Engineering,* **24**, 568.

Leeman, V., Earing, D., Vincent, R.K. and Ladd, S. (1971), *NASA Earth Resources Spectral Information System: a data compilation,* NASA CR-31650-24-T.

Meyer, M.P. and Maklin, H.A. (1969), 'Photointerpretation techniques for Ektachrome I.R. transparencies', *Photogrammetric Engineering,* **35**, 1111.

Mintzer, O.W. (1968), 'Soils' in *Manual of Color Aerial Photography,* Smith, J. (ed.), American Society of Photogrammetry.

Mott, P.G. (1966), 'Some aspects of colour aerial photography in practice and its applications', *Photogrammetric Record,* **5**, 221.

Myers, V.I. and Heilman, M.D. (1969), 'Thermal infrared for soil temperature studies', *Photogrammetric Engineering,* **10**, 1024.

NASA, (1973), *Symposium on significant results from the ERTS-1, Vol. I and II, NASA,*

Parks, W.L. and Bodenheimer, R.E. (1973), 'Delineation of major soil associations using ERTS-1 imagery, *Symposium on significant results from ERTS-1,* pp. 121-125.

Parry, J.T., Cowan, W.R. and Heginbottom, J.A. (1969), 'Soil studies using color photos', *Photogrammetric Engineering,* **35**, 44.

Pomerening, J.A. and Cline, M.G. (1953), 'The accuracy of soil maps prepared by various methods that use aerial photographic interpretation', *Photogrammetric Engineering,* **19**, 809.

Romanova, M.A. (1968), 'Spectral luminance of sand deposits as a tool in land evaluation', in *Land Evaluation*, Stewart, G.A., (Ed.), Macmillan, Australia, pp. 342-348.

Roscoe, J.H. (1955), (Moderator) Panel — on Photointerpretation, *Photogrammetric Engineering*, **21**, 564.

Sager, R.C. (1951), 'Aerial analysis of permanently frozen ground', *Photogrammetric Engineering*, **17**, 551.

Seevers, P.M. and Drew, J.V. (1973), 'Evaluation of ERTS-1 imagery in mapping and managing soil and range resources in the sand hills region of Nebraska', *Symposium on significant Results from ERTS-1*, NASA, pp. 87-95.

Simakova, M.S. (1964), 'Soil Mapping by Colour Aerial Photography', Israel Programme for Scientific Translations, Jerusalem.

Smith, H.T.V. (1943), *Aerial Photographs and their Applications*, Appleton-Century Crofts, New York.

Speight, J.G. (1968), 'Parametric description of land form', in *Land Evaluation*, Stewart, J.A. Macmillan, Australia, pp. 239-250.

Thomson, F.J. and Roller, E.G. (1973), 'Terrain classification maps of Yellowstone National Park, *Symposium on significant results from ERTS-1*, pp. 1091-1095.

Way, D.S. (1973), *Terrain Analysis*, Dowden, Hutchinson and Ross, Stroudsburg.

Webster, R. and Wong, I.F.T. (1969), 'A numerical procedure for testing soil boundaries interpreted from air photographs', *Photogrammetria*, **24**, 59.

Welch, R. (1966), 'A comparison of aerial films in the study of the Breidamerkur glacier area, Iceland, *Photogrammetric Record*, **5**, 289.

Werner, H.D., Schmer, F.A., Horton, M.L., and Waltz, F.A. (1973), *Application of Remote Sensing Techniques to Monitoring Soil Moisture*, Environmental Research Institute of Michigan, pp. 1245-1258.

Weston, F.C. and Myers, V.I. (1973), 'Identification of soil associations in Western South Dakota on ERTS-1 imagery', *Symposium on significant results from ERTS-1, NASA*, pp. 965-972.

13 Rock and mineral resources

13.1 The use of air photo-interpretation in studies of rocks and minerals

The use of remote sensing techniques in geology is long established through the use of aerial photography in photo-geologic studies. Several text books are available which deal with the methods used in photo-geology and some sources are suggested at the end of this chapter.

Aerial photographs have provided a great deal of data for geological studies in all parts of the world. They can be interpreted to give information on (i) structure (ii) lithology of rocks. In the study of structural geology such features as bedding, dip, foliation, folding, faulting, and jointing can be observed. Aerial photographs provide evidence of bedding through the occurrence of ridges in the stereomodel and differences in tonal response where beds differ in their mineral constituents. Sometimes certain beds can be recognised by a constant lithological interface which is so distinctive that they can be regarded as marker beds or marker horizons. The dip slopes of rocks can often be recognised more reliably on a stereomodel than on the ground because of the synoptic view of an area of dipping sediments obtained from the air. Accurate measurements of dip can be made by photogrammetric measurements on stereopairs. In regional mapping, however, geologists normally assess the dip by eye and place the dip into categories (e.g. $< 10^0$, $10-25^0$, $25-45^0$, $> 45^0$, $< 90^0$). (Plate 13.1).

Any discussion of foliation in rocks is made more difficult by the fact that the term foliation is used in different senses in Britain and America. British geologists distinguish between schistosity and foliation so that for them segregation of minerals into thin layers or folia is foliation whereas parallel orientation of such minerals is referred to as schistosity. On the other hand American geologists give the term a wider meaning so that lithological layering, preferred dimensional orientation of mineral grains and surfaces of physical discontinuity and fissility resulting from localized slip may be included in the term. In this chapter the American usage of the term will be adopted. Lineaments resulting from foliation are generally parallel to one another but are short and do not normally consist of long continuous ridges or valleys (Plate 13.2).

Folding in rocks is often apparent on air photographs where it would not be noticed in the field by a ground surveyor. The geologist can normally see a representation of the whole fold in the stereomodel so that the axis and plunge of a fold structure can

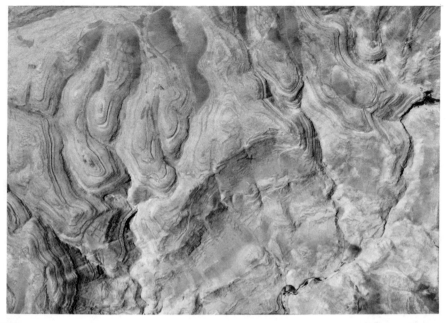

Plate 13.1
An area of gently dipping sediments occurring in Jordan. Note how beds of differeing lithology are characterized by different tones on the photograph. Field systems can be seen on the gentler slopes. (Courtesy, Hunting Surveys Limited, Boreham Wood, Herts.)

often be seen very readily (Plate 13.3).

Faulting in rocks occurs where there has been a fracture along which rocks have slipped relative to one another. Generally faults provide fairly straight features and since the fracture is a zone of weathering and weakness the surface manifestation is often one of negative relief. Where mineralization has taken place along a fault line, however, there may be positive rather than negative surface features. In most instances the most reliable evidence of faulting is displacement of bedding along negative linear surface features (Plate 13.4).

Joints from patterns in rocks which are very similar in photographic appearance to faults i.e. they often provide fairly straight negative features. The distinction between joints and faults can sometimes be made by careful observation to see if relative movement has taken place in the beds. If relative movement can be seen then the feature can be classified as a fault, and conversely if no movement can be detected it is better to record the feature as a joint. Jointing often plays a part in determining the patterns of river networks (Plate 13.5). It can also be associated with topographic forms characteristic of granite areas and in such cases the boundaries of granite intrusions can often be plotted with some accuracy.

Whereas structural geology can often be interpreted with considerable certainty

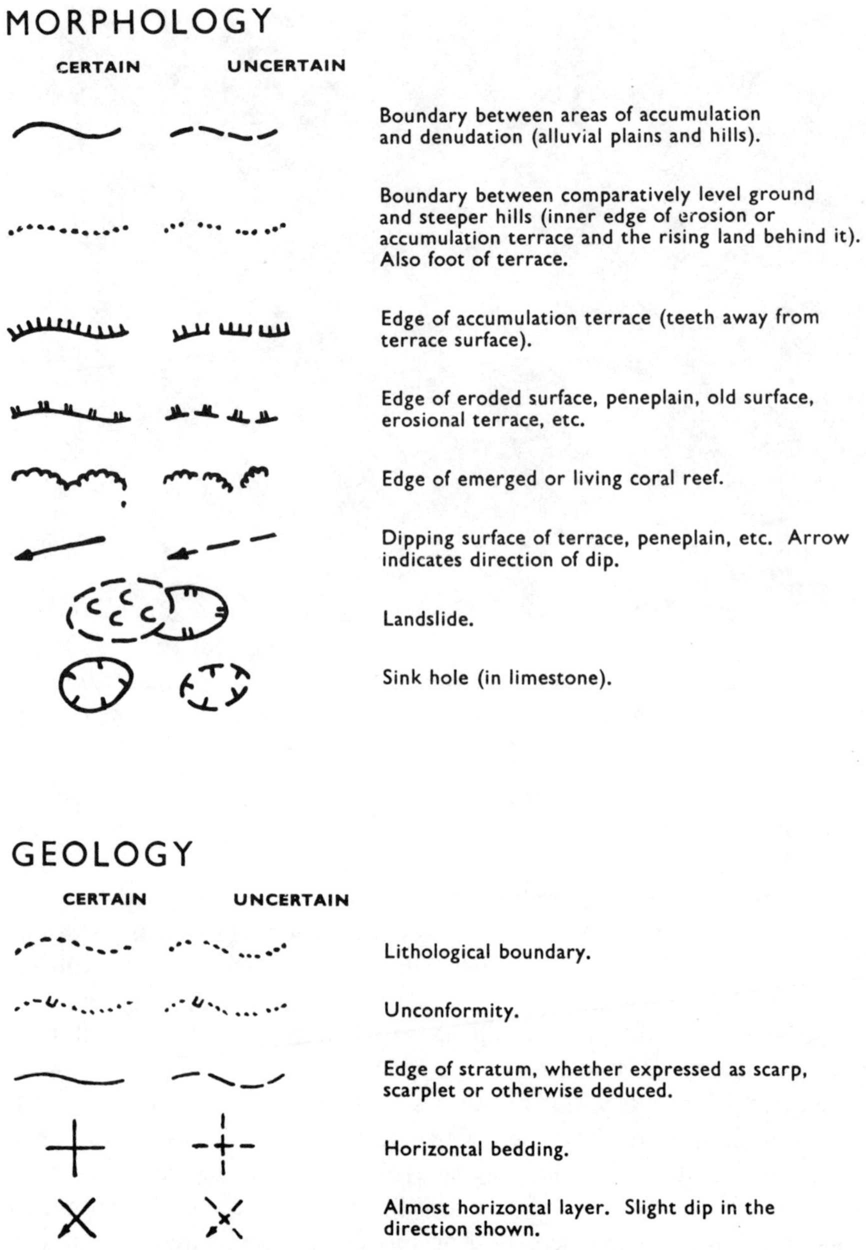

MORPHOLOGY

CERTAIN   **UNCERTAIN**

Boundary between areas of accumulation and denudation (alluvial plains and hills).

Boundary between comparatively level ground and steeper hills (inner edge of erosion or accumulation terrace and the rising land behind it). Also foot of terrace.

Edge of accumulation terrace (teeth away from terrace surface).

Edge of eroded surface, peneplain, old surface, erosional terrace, etc.

Edge of emerged or living coral reef.

Dipping surface of terrace, peneplain, etc. Arrow indicates direction of dip.

Landslide.

Sink hole (in limestone).

GEOLOGY

CERTAIN   **UNCERTAIN**

Lithological boundary.

Unconformity.

Edge of stratum, whether expressed as scarp, scarplet or otherwise deduced.

Horizontal bedding.

Almost horizontal layer. Slight dip in the direction shown.

Fig. 13.1
Symbols for photo-interpretation in geology. (Source: Shell Petroleum Company Limited)

GEOLOGY (continued)

CERTAIN	UNCERTAIN	
		Gentle dip.
		idem, with scarp (edge of bed).
		Moderately steep dip.
		Steep dip.
		Moderately steep dip, when length of dip slope too small for arrow.
		Steep dip, idem.
		Comparatively hard vertical layer: vertical dip.
		Overturned beds.
		Linear feature of uncertain or unknown origin.
		Axis of anticline.
		idem, showing pitch of anticline.
		Axial culmination of anticline.
		Axis of syncline.
		idem, showing pitch of syncline.
		Axial depression of syncline.
		Fault, with indication of throw when possible.
		Fault zone.

Plate 13.2
*Oblique aerial photograph of the Fraser River area, Canada. (56° 45' N, 63° 35' W). Note the linea-
ments in the surface which are made easily visible by the snow collecting in the lineament depressions.
Both short and long lineament features are displayed.* (Courtesy Royal Canadian Air Force.)

from aerial photographs, lithological interpretation is less easy. If the researcher is
familiar with a particular field area and has used air photos extensively, however, it
may be possible to recognize rock type with considerable accuracy. Where an attempt
is made to recognize rock types in unfamiliar areas from photo features alone, the
task is much more difficult and uncertain. Various strands of evidence must then be
used — in other words the principle of convergence of evidence must be applied.
It has been suggested that the following stages may be included in interpretation of

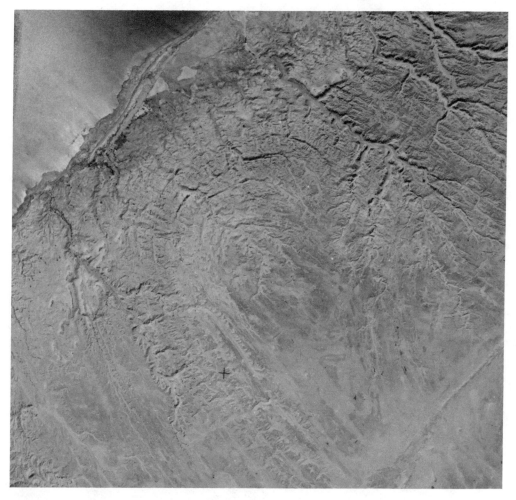

Plate 13.3
A plunging fold structure near the Dead Sea, Jordan. (Courtesy, Hunting Surveys Ltd., Boreham, Wood, Herts.)

lithology and structure:
(a) The recognition of the climatic environment, e.g. temperate, tropical rain-forest, savannah, desert, etc.
(b) The recognition of the erosional environment, e.g. very active, active, inactive.
(c) The recognition and annotation of the bedding traces of the sediments or meta-sediments (altered sediments).
(d) The recognition and delineation of areas of outcrop that do not indicate bedding e.g. intrusions in horizontally bedded rocks.

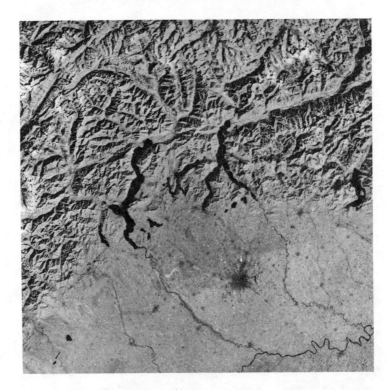

Plate 13.4
Faulting in Alpine Structures fringing the North Italian plain as seen in ERTS-1 imagery obtained in Band 7, 7th October 1972. (From Boriani, Marino and Sacchi, 1974)

Plate 13.5
Faulted terrain in an arid environment in Jordan. Note the sharp changes of direction in the wadi system resulting from the effects of faulting and jointing. (Courtesy, Hunting Surveys Ltd., Boreham Wood, Herts.)

(e) The recognition and delineation of areas of superficial cover that do not indicate bedding.

(f) The restudy of the bedding traces around the noses of folds to determine if possible the approximate position of the axes of folds.

(g) The study of lineaments transverse to the bedding traces to determine whether they represent faults, dykes, joints, or combinations of these.

It is important to use a recognized set of symbols for annotation of prints in geological studies so that work can be interpreted by co-workers. In Figs. 13.1 and 13.2 the symbols used for photo-geological work by the Royal Dutch/Shell Group are illustrated. An example of a restricted working legend used by the Overseas Division of the Institute of Geological Sciences, London, is shown in Fig. 13.3. This set of symbols was used in the interpretation of the geological structure of an area of folded sediments in Australia (Plate 13.6) for which the final photo-interpretation is shown in Fig. 13.4.

It has long been recognized that certain plants are characteristic of soils which are found over rocks containing particular minerals. These plants are termed indicator plants and they have been used as guides by mineral prospectors seeking the occurrence of ore bodies. There is now a considerable literature dealing with the use of geobotanical techniques for mineral exploration and since vegetation boundaries are often well displayed on air photographs attempts have been made to trace indicator plants on the photographs and thereby outline potential mining areas. These methods have been successfully applied in Africa and Australia and the development of multi-spectral photography together with infrared colour photography has made such geobotanical studies of growing interest.

The literature dealing with the use of air photography in the study of rocks and mineral resources is now voluminous and most major exploration companies rely heavily on photogeological work. There are also separate photogeological departments in various governmental institutes e.g. Institute of Geological Sciences, Britain. The reader is directed towards bibliographies on air photo interpretation in geology in the references at the end of this chapter.

13.2. The use of non-photographic sensors for studies of rocks and minerals

13.2.1 Thermal infrared scanning

The development of non-photographic sensors has provided remote sensing data in the infrared (thermal) and microwave regions of the spectrum which are potentially very important for geological studies. Infrared scanning has been used for studies of Italian volcanoes and volcanic deposits. These have been overflown with a two-channel Daedalus thermal scanner. The ratio of two thermal channels when plotted can be considered as a relative emissivity map of volcanic materials. It has also been found

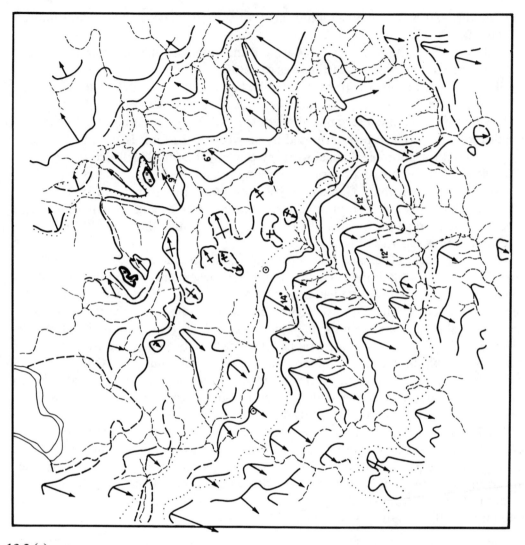

Fig. 13.2 (a)
Photogeological interpretation of Tropical Rain Forest areas in New Guinea. Scale of original 1: 40 000.
(Source: Shell Petroleum Company Limited)

that such a ratio method can be used for mapping the texture of volcanic materials.

In the Soviet Union airborne infrared images have been used to study sand desert features in the Repetek region. Landscape features such as barchan ridges and sand hills have been found to show different thermal patterns according to differences in slope steepness, density and composition of the vegetation cover, solar elevation and wind speed and direction. The temperature graduations were found to have a large range

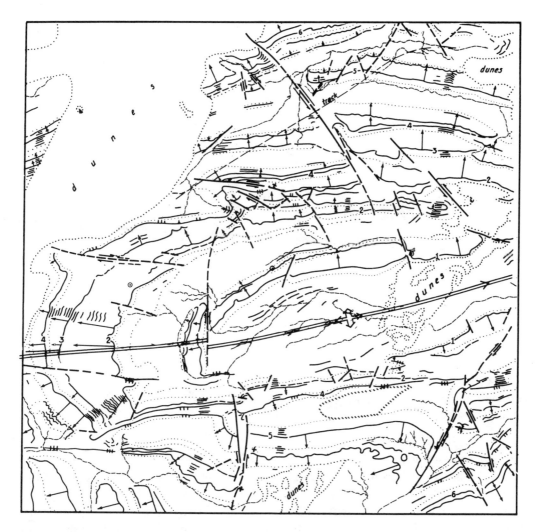

Fig. 13.2 (b)
Photogeogical interpretation of North African Desert region, Algeria. Scale of original 1: 33 000.
(Source: Shell Petroleum Company Limited.)

from 55° C on the sunlit slopes of barchans to 16° C in the shade of desert bushes.

Another application of thermal scanning arises from the fact that mineral prospectors have long recognized that there are differences in soil temperature over ore bodies, therefore thermal infrared line scan techniques are potentially useful for mineral exploration. An example of this is shown in Fig. 13.5 where the microdensitometer scan across the Dugald River Lode in Australia indicates that it is marked by a fairly clearly defined zone of high emission in line scan imagery.

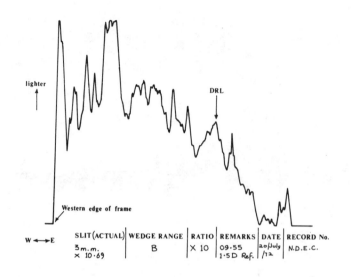

Fig. 13.3
Microdensitometer scan line of the infrared line scan imagery of the Dugald River Lode area.
(Source: Custance, 1974.)

13.2.2 Microwave radar

Perhaps the most extensively used non-photographic system for geologic studies is the microwave radar system. Side-looking radar systems are particularly valuable in equatorial and maritime regions where persistent cloud renders photography difficult to acquire. Furthermore the radar system illuminates the ground from the side so that shadowing effects are produced similar to those of low-sun photography (Fig. 4.8). These are well shown in the radar imagery of the Malvern Hills illustrated in Plate 4.5 (a).

Shape, pattern, tone and texture are used in the interpretation of radar imagery but the side illumination of radar systems places an emphasis on patterns i.e. lineaments. Tones may vary greatly according to look angle and so must be used with caution. Textures in the imagery mainly reflect the physical form i.e. roughness of the surface.

Lineaments are often well displayed in radar images (Plate 11.3) and regional fault and fold patterns can be conspicuously displayed and readily mapped. The orientation of the radar system can, however, give rise to bias in the representation of linears. Also where areas are in the radar shadow (see Plate 13.8) ground information is lost.

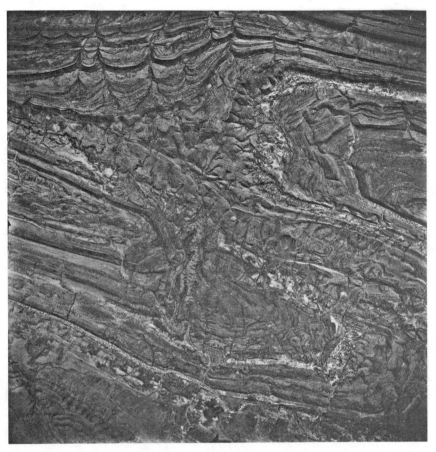

Plate 13.6
Folded sediments in Australia. Note the pitching anticline on the left of the figure and the dipping sediments on the right. The photo-geological map of this area is shown in Fig. 13.4. (Courtesy, Dr. E.A. Stephens, Institute of Geological Sciences.)

13.3 Space observations for the study of rocks and mineral resources

The advent of the satellite platforms has provided the geologist with a new and wider perspective on geological structures. Thus photographs of New Mexico obtained from the Apollo-Saturn 6 spacecraft launched in April 1968 were examined with interest by geologists. They found a major northeast-trending lineament which was not shown on the most recent gelologic map available for the New Mexico area (Fig. 13.5 (a). However, the major impact of the satellite platform has come with the ERTS-1 satellite with its MSS data. Much geological work has been achieved with this data and some selected examples will be discussed below. The general achievements of the Landsat programme

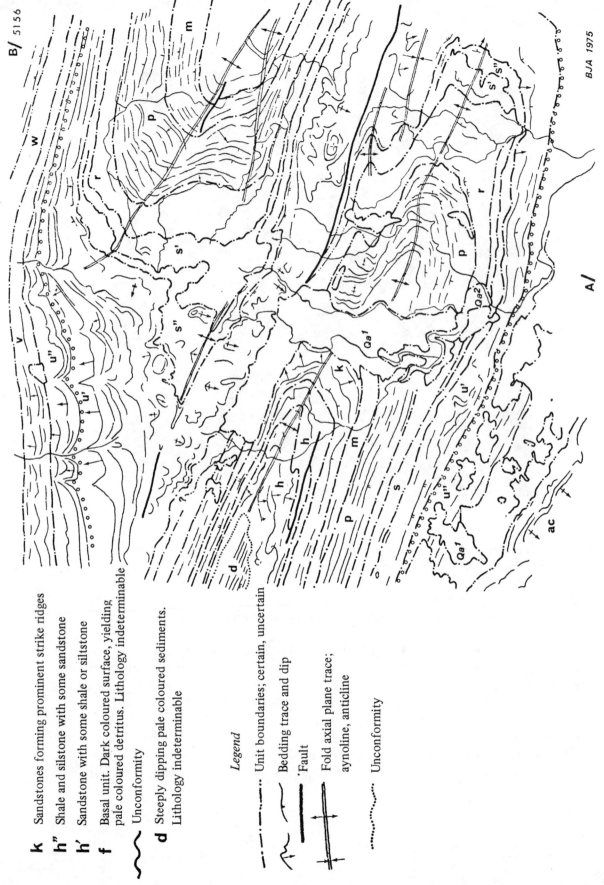

BJA 1975

k   Sandstones forming prominent strike ridges

h"   Shale and silstone with some sandstone

h'   Sandstone with some shale or siltstone

f   Basal unit. Dark coloured surface, yielding pale coloured detritus. Lithology indeterminable

Unconformity

d   Steeply dipping pale coloured sediments. Lithology indeterminable

Legend

— ·· — ··   Unit boundaries; certain, uncertain

Bedding trace and dip

Fault

Fold axial plane trace; aynoline, anticline

··········   Unconformity

Fig. 13.4

The structural map prepared from a vertical photograph of folded sediments in Australia. (Source: Institute of Geological Sciences, London.)

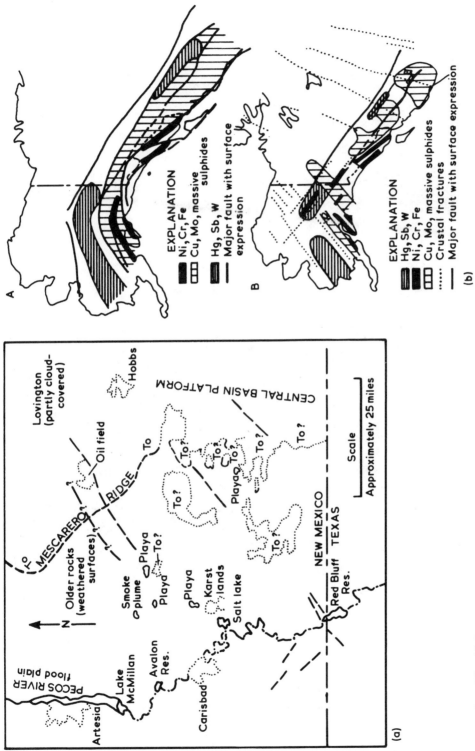

Fig. 13.5

(a) An example of linear features in the New Mexico area mapped from an Apollo photograph obtained in 1968. (Source: Carter and Meyer, 1969).

(b) Geologic conditions in Alaska interpreted from Nimbus IV data. (A) Conventional concept of the lithologic belts (B) Alternative linears seen on the Nimbus image. (Source: Lathram et al., 1973)

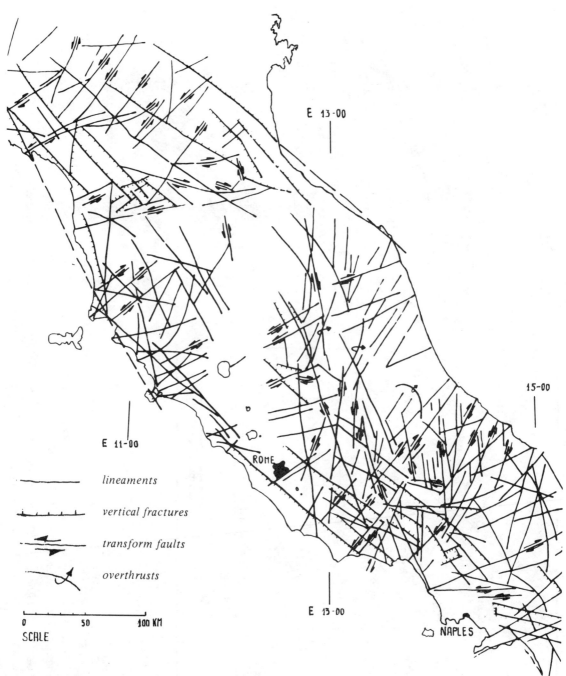

Fig. 13.6
Main fault systems in central Italy, as interpreted from ERTS-1 data. (Source: Bodechtel, J. *et al.,*
1974, Proc. Symp. *European Earth Resources Satellite Experiments,* ESRO SP–100, 205–215)

have been summarised by ERTS investigators. The most significant results can be listed as follows:

(a) The broadest use of ERTS imagery in geologic mapping lies in the construction of regional structural (or tectonic) maps. Major geologic features, contact between distinctly different, thick rock units, and land form types can be effectively mapped from ERTS images with details comparable to and sometimes superior to mapping by conventional aerial photo or ground methods.

(b) Computer produced geologic maps at scales up to 1:24 000 can be made from ERTS images with surprisingly good accuracy, providing training set data and other supervised methods are applied.

(c) Winter imagery, with foliage-free scenes or snow-cover enhancements, provides useful data for mapping in many instances. Such imagery is especially valuable in seasonally vegetated areas and areas with considerable relief.

A number of applications for ERTS-1 geologic studies have been identified. In regions for which only poor quality or outdated maps have been produced ERTS data offers a method of constructing good general maps at small scales. Existing maps can also be checked against ERTS imagery to correct mislocated or omitted rock unit contacts, geologic structures (fold axes for example) and lava flows. ERTS-generated maps depicting structural information (particularly lineaments) also have utility in the search for ore deposits, oil accumulations and ground water zones.

It must be borne in mind, however, that field checking has often been insufficient to verify the accuracy and correctness of ERTS mapped data. Also it would appear that no reliable identifications of lithologic types have been consistently made. The band ratio techniques (e.g. MSS 7/5) lead to some enhancement of visual images that improves the separation of rock types. However, the ranges of ratio values for most common rocks and minerals overlap. Unique spectral signatures have not been found to occur within the MSS sensing wavelengths. In fact the ratio interval for haematite and serpentine—strikingly different minerals— is almost identical.

It is also found that individual stratigraphic units are usually thinner than the linear resolution capability of ERTS-1 (70–100 m). Thus remote sensing 'units' are often groupings of stratigraphic units and they sometimes have limited applicability to standard geologic mapping procedures.

With these advantages and limitations of ERTS imagery in mind one can examine some selected examples of geologic interpretations in order to illustrate some of the techniques employed.

An interesting application of the mapping of linears and faults is provided by workers comparing Nimbus and ERTS imagery. Prior to launch of ERTS-1 a cloud free image of Alaska was obtained by the Nimbus IV Image Dissector Camera System. This image showed a set of northwest and northeast trending linears that suggested previously unrecognized geologic structures deep in the Earth's crust.

Linears and faults on ERTS-1 images have corroborated and added detail to the

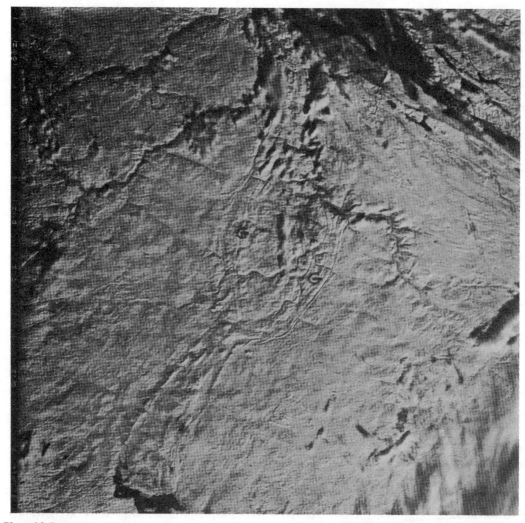

Plate 13.7
ERTS-1 image; early winter, October 9th 1972; sun elevation 25°; no geometric correction; MSS, band 6, of New England, U.S.A. Linears of quartzite marked G. (Source: Gregory, A.F., 1973.)

initial geological trends noted on the Nimbus imagery. These linears have been compared to the relation of known mineral deposits and fundamental fractures in the Canadian Cordillera. As a result it has been possible to provide an alternative map to guide mineral resource application in Alaska and Western Canada. (Fig. 13.5b). The great detail obtainable from ERTS images is also borne out by the detailed mapping of main fault systems revealed by ERTS-1 data for the Italian peninsula as shown in Fig. 13.6.

Many researchers have noted that snow cover does not obliterate linears, indeed

1014-17375-3-5 06AUG72 N34-35 W112-37 RED .6-.7
COMCAT - STRETCH
HORIZONTAL BAR FILTER
FILTER - SKEW
MUSK

Plate 13.8
Computer-enhanced product of the Verde valley region, Central Arizona. The horizontal bar filter removes banding and enhances small linear features in other directions. (Source: Goetz A.F.H. *et al.*, 1973.)

major structural features may be accentuated (Plate 13.7). Observations in
New England, U.S.A. indicate that a heavy blanket of snow (9 in/22.5 cm)
accentuates major structural features whereas a light dusting of snow (1 in/2.5 cm)
accentuates more subtle topographic expressions. Comparisons of snow free and snow
covered ERTS-1 images for the same area have shown that snow covered imagery
allows more rapid fracture analysis and provides additional fracture detail.

Some researchers have used the methods of spectral ratioing of reflected radiances
of selected pairs of ERTS multispectral channels and production of analog ratio
images for mapping of large exposures of iron compounds in Wyoming. The data was
combined with laboratory data as training sets wherever possible.

Enhancement techniques were employed in which the digital data of one channel
was divided by the data from a second channel. For example where channel 7 data
was divided by channel 5 data the resultant ratio digital graymap is given the notation
R_{75}. When the analog ratio images were printed in colour it was found that green
areas represented vegetation, violet areas primarily rock and vegetation, blue areas rock
outcrop and red represented iron rich outcrops. Darkest red was found to occur only
in an iron mine area and along pond edges, where muds and tailings were present. Other
images were created by analog ratio images. For example the analog R_{74} showed iron
mines as unique dark areas in the scene. These digital ratio and analog ratio maps can
be used to focus the attention of the geologist or scientist on areas in the scene where
chemical properties are different from those of adjoining or surrounding localities.

Enhancement of ERTS images can also be achieved by a process of directional
filtering. This can be used to remove horizontal banding in the scene and enhance small
oriented features. An example of such directional filtering is shown in Plate 13.8 for an
area of the Verde valley region of Central Arizona.

Automatic pattern recognition of the LARSYS type described on p. 275 can, of
course, be applied to geological materials. However it is necessary to make certain
assumptions if such mapping is to be undertaken. Firstly, it must be assumed that
subsurface materials will manifest themselves as spectrally separable classes at the Earth's
surface. Since subsurface rocks are normally veneered by soil and vegetation it must
also be assumed that spectrally separable surface features are correlated with subsurface
variations.

The second major assumption is that lithologic types are naturally segregated into
a limited number of discrete compositional and textural categories which can be recognized in some classification system. This assumption is clearly wrong, as it is for
many soils. However it does provide a basis for grouping similar lithologies into discrete classes. Where intergrades occur e.g. sandy-shale the classifier may assign to either
sandstone or shale in a random fashion.

Where these automatic recognition techniques have been applied to ERTS images
of southwest Colorado a correlation of 89.8 per cent has been obtained with existing
geologic maps. It is apparent, however, that extensive ground truth observations and

aircraft underflight data are needed before definitive statements can be made concerning the reliability and accuracy of machine made maps. Nevertheless the vast amount of information now coming from satellite sensors makes it necessary to think in terms of automatic mapping procedures for the future.

References

Allum, J.A.E. (1966), *Photogeology and Regional Mapping,* Pergamon, London.

Billingsley, F.C. and Goetz, A.F.H. (1973), 'Computer techniques used for some enhancements of ERTS images', *Symposium on significant results from ERTS-1,* NASA, pp. 1159-1167.

Blodget, H.W. and Anderson, A.T. (1973), 'A comparison of Gemini and ERTS imagery obtained over Southern Morocco', *Symposium on significant results from ERTS-1,* NASA, pp. 265-272.

Boriani, A., Marino, C.M. and Sacchi, R. (1974), 'Geological features on ERTS-1 images of a test area in West-Central Alps', in *Proc. Symp. on European Earth Resources Satellite Experiments,* ESRO SP-100.

Carter, W.D. and Meyer, R.F. (1969), 'Geologic analysis of a multi-spectrally processed Apollo space photograph', *New Horizons in Colour Aerial Photography.* American Society of Photogrammetry, pp. 59-64.

Cassinis, R., Lechi, G.M. and Tonelli, A.M. (1974), 'Contribution of space platforms to a ground and airborne remote sensing programme over active Italian volcanoes', *Proc. Symp. on European Earth Resources Satellite Experiments,* ESRO SP-100, Paris, pp. 185-197.

Cobb, G.C. (1943), 'Bibliography on the interpretation of aerial photographs and recent bibliographies on aerial photography and related subjects', *Bull. Geol. Soc. Amer.,* **54,** 1195.

Cole, M.M. (1973), 'Geobotanical and biogeochemical investigations in the sclerophyllous woodland and scrub associations of the Eastern Goldfields area of Western Australia', *J. Appl. Ecol.,* **10,** 269.

Cole, M.M., Owen-Jones, E.S. and Custance, N.D.E. (1974), 'Remote Sensing in mineral exploration', in *Environmental Remote Sensing; applications and achievements,* Barrett, E.C. and Curtis, L.F. (eds.), Edward Arnold, London, pp. 49-66.

Froelich, A.J. and Kleinkampl, F.J. (1960), 'Botanical prospecting for uranium in the Deer Flat area, White Canyon District, San Juan County, Utah', *U.S. Geological Survey Bulletin,* 1085-B.

Goetz, A.F.H., Billingsley, F.C., Elston, D., Luchitta, I., and Shoemaker, E.M. (1973), 'Preliminary geologic investigations in the Colorado Plateau using enhanced ERTS images'. *Symposium on significant results obtained from ERTS-1,* NASA, pp. 403-411.

Gregory, A.F., (1973), 'Preliminary assessment of geological applications of ERTS-1 imagery from selected areas of the Canadian Artic, *Symposium on significant results from ERTS-1,* NASA, pp. 329-344.

Lathram, E.H., Taillevr, I.L. and Patton, W.W. (1973), 'Preliminary geologic application of ERTS-1 imagery in Alaska', *Symposium on significant results from ERTS-1,* NASA, pp. 257-264.

Martin-Kaye, P. (1974), 'Application of side looking radar in earth-resource surveys', in *Environmental Remote Sensing: applications and achievements,* Barrett, E.C. and Curtis, L.F. (eds.), Edward Arnold, London, pp. 29-48.

Melhorn, W.N. and Sinnocks, S. (1973), 'Recognition of surface lithologic and topographic patterns in Southwest Colorado with ADP techniques, *Symposium on significant results from ERTS-1,* NASA, pp. 473-481.

Miller, V.C. (1961), *Photogeology,* McGraw Hill, New York.

Ray, R.G. (1960), *Aerial Photographs in Geologic Interpretation and Mapping,* U.S. Geological Survey, Professional Paper 373.

Short, N.M. (1973), 'Mineral resources, geological structure and landform surveys', *Symposium on significant results obtained from ERTS-1,* Vol. III, NASA, pp. 30-46.

Vincent, R.K. (1973), 'Ratio maps of iron ore deposits, Atlantic City District, Wyoming, Symposium *on significant results from ERTS-1,* NASA, pp. 379-386.

Vinogradov, B.V., Grigoryev, A.A., Lipatov, V.B. and Chernenko, A.P. (1972), 'Thermal structure of the sand desert from the data of IR aerophotography', *Proceedings of the 8th International Symposium Remote Sensing of Environment,* Ann Arbor, Michigan, pp. 729-737.

Wobber, F.J. and Martin, K.R. (1973), 'Exploitation of ERTS-1 imagery utilising snow enhancement techniques', *Symposium on significant results from ERTS-1,* NASA, pp. 345-351.

 **Crops and
land use**

14.1 Introduction

One of the most pressing problems facing society today is that of feeding and housing
the people of the world. This means that land should be used to its full potential. In
order to achieve such an aim it is necessary to have up to date information concerning present land use, and changes in land and crop condition.

At the present time agriculture dominates the land use of most countries. For example
in 1965 some 81.5 per cent of the land in the United Kingdom was devoted to agriculture. However, there are now competing claims for land for recreation, forestry and
urban growth. From present trends it is possible to forecast changes in land use likely
to occur in the United Kingdom by the year 2000 (Table 14.1).

Although land use maps have been made for various parts of the world these maps
are usually out of date by the time they are published. In many of the developing
countries there are no land use maps available at scales which are useful for planning
purposes. There is, therefore, a great need for remote sensing techniques capable of
identifying land use.

14.2 Photographic studies of crops and land use

14.2.1 Introduction
Early examples of land use studies using aerial photography include experiments in
land classification in Northern Rhodesia and in the 1940's there were air photo inter-

Table 14.1
Land use in the United Kingdom (estimated).

Land use	1965		2000	
	Area 10³ hectares	% Total	Area 10³ hectares	% Total
Agricultural land	19 624	81.5	18 122	75.25
Urban land	2 043	8.49	2 745	11.43
Forestry and woodland	1 817	7.54	2 617	10.86

Plate 14.1a
Yellow rust infection in Winter Wheat. The original foci are shown in this infrared photograph as dark patches. (Courtesy, Ministry of Agriculture, Crown copyright reserved.)

pretation studies of agriculture and forestry in many areas of the world. These early studies were, of course, limited to the use of panchromatic photography. Thus photographic features such as patterns of fields, tone, texture, shape and size were used to aid interpretation. Where photographs were available for different times of the year changes in the photographic image might be caused by tillage, crop growth, crop rotation or conservation measures. Photo interpretation keys were developed for various crops for particular agricultural regions. A useful review of early techniques in mapping agricultural land use from air photos is given in the Manual of Photographic Interpretation published by the American Society for Photogrammetry, 1960.

These early studies were, however, restricted to the use of black and white photography. It was not until the development of colour and infrared colour photography that new horizons opened up in respect of mapping crops and land use.

,Similarly the potential usefulness of aerial photography for the detection of crop diseases (see Plate 14.1) was recognised by British and American research workers at an early stage. However, the development of colour, infrared colour and multi-spectral photography has added further impetus to this area of research in the last

Plate 14.1b
Potato blight infection shown by infrared photography. The infection probably originated near the margin of the field at top centre where a large dark patch shows a major focus. It may have originated from a clamp or dump of discarded potatoes close to this spot. The small foci show 'tails' indicating the spread of infection downwind. (Courtesy, Ministry of Agriculture, Crown copyright reserved.)

decade. Diseases such as potato blight (*Phytophthora infestants*), take-all disease (*Ophiobolus graminis*) in cereals and leaf spot (*Cercospora*) in sugar but have been detected and monitored by the use of panchromatic, colour and infrared colour photographs.

Where crops have died from disease or adverse environmental conditions the dead portions of the crop can be readily identified using infrared colour photography. Healthy plants image in red on this emulsion whereas dead portions normally image in greyish or blue/green colours. In the case of diseased plants the colours of diseased plants often range from salmon pink (slight disease) to dark brown (severe disease).

Crop discrimination with colour infrared photography has been studied closely and it has been found that the time of year, location and environment affect the percentage accuracy obtainable when discriminating crop types. Generally speaking crop discrimination in late July in northeastern Kansas was difficult because all crops were imaging in red at that time. Studies in Britain recently suggest, however, that quality can be distinguished on colour infrared photographs of pasture. (See colour plates).

14.2.2 Multispectral techniques

The development of multispectral photography and multispectral scanning techniques has provided an added dimension to remote sensing techniques in crop and land use studies. The principal objective now being pursued is that of automatic recognition of land use categories and crops by means of computer analyses of spectral reflectance

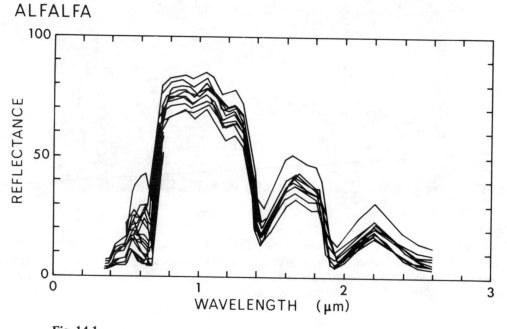
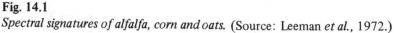

Fig. 14.1
Spectral signatures of alfalfa, corn and oats. (Source: Leeman *et al.*, 1972.)

data. Two approaches can be adopted in mapping from spectral data. On the one hand a training set of data can be obtained and used to control the computer processing of the remote sensing data. For example the spectral reflectance characteristics (density response at different wavelengths) of a crop may be determined by laboratory and

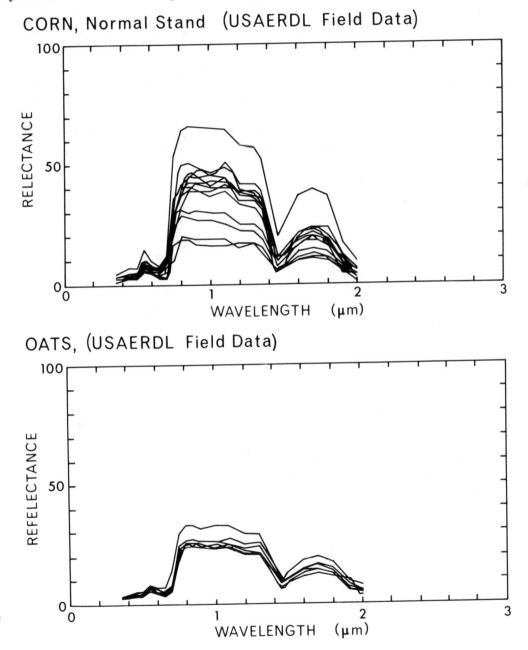

field measurements. The data can then be used to supervize the classification of the remote sensing data. Alternatively, it is possible to adopt an unsupervised classification approach. In this case the remote sensing data is grouped by clustering algorithms.

The concept of the spectral signature lies at the heart of the multispectral approach and a great deal of effort has been made to measure the spectral reflectance characteristics of a number of crops. A major source of data concerning crop signatures is that obtained under the Air Force Target Signatures Measurement Program in the U.S.A. These were mostly measurements in the visible wavelengths and much of the data is derived from laboratory investigations. Examples of the signatures obtained for certain crops are given in Figs. 14.1a,b,c.

14.2.3 Ground truth for crop and land use studies

Useful though laboratory measurements may be the spectral characteristics of a crop when sensed in the field may be greatly modified by factors such as crop geometry, per cent crop cover, crop colour, crop moisture, disease or fertilizer patterns, grazing patterns and husbandry patterns. Therefore there is an increasing demand for field data as well as laboratory data for the calibration of sensors and checking of data processing techniques. Such ground data is normally referred to as 'ground truth'.

The range of ground truth data required varies according to the farming region and its soil, relief and climatic conditions. Furthermore, the cost of acquisition of remote sensing imagery often determines that it be used for more than one purpose, e.g., in the agricultural context for crop recognition, soil drainage mapping and land quality evaluation. Generally, four categories of data are required for multipurpose land use studies in rural areas as follows:

(a) Site morphology.

(b) Crop/vegetation cover characteristics.

(c) Cultivation/husbandry features.

(d) Soil surface conditions.

A specimen data collection form devised for data in categories (a)–(c) above is shown in Fig. 14.2. Soil surface conditions can normally be recorded separately using established methods described in soil survey manuals.

Where large quantities of ground data are to be collected and handled it is necessary to develop a system of computer storage. In these circumstances coding of data in a computer compatible form becomes desirable. Such coding is relatively straightforward where a limited range of data is to be recorded and where class intervals are known or can be predicted. In our experience, however, it is difficult to provide unambiguous codes which will cover all, or even most, land use conditions of possible significance to imagery evaluation. There is a real risk that different ground surveyors will code the same conditions differently. Attempts to devise comprehensive coding systems can, therefore, lead to complex systems which impair speed and efficiency in field surveys. Thus, where coding is employed, experimentation is necessary to test the coding system and to train field staff.

DATE . . . .   LAND USE * . . . .   FIELD REF . . . .

CROP CONDITIONS   *SOIL CONDITIONS*

Stage   * [   ]   % Soil exposed   [   ]

Height: average   [   ]   Surface   general   [   ]
          range   [   ]   colour: pattern   * [   ]
          pattern   * [   ]   extent   * [   ]
          extent   [   ]   comment   [   ]
          comment   [   ]

Colour: general   * [   ]   Roughness: furrowed   [   ]
          pattern   * [   ]   normal tilth [   ]
          extent   * [   ]   cloddy   [   ]
          comment   [   ]   panned   [   ]

Disease: type   [   ]   Surface   abundance   [   ]
          extent   * [   ]   stones: %   [   ]
          comment   [   ]   size   [   ]
                                                       type   [   ]

Weeds: species   [   ]
          density   * [   ]   Surface   * [   ]
          comment   [   ]   moisture

Husbandry   *Site
          Ploughed   [   ]   Morphology:*
          Harrowed   [   ]   Gradient   * [   ]
          Drilled   [   ]   Slope type * [   ]
          Rolled   [   ]
          Wheelings   [   ]   Microrelief   [   ]
                                                       type
          Grazing   [   ]   extent   * [   ]
            method   [   ]   Field   *
          Livestock   Boundary   *

                        * Codes available (e.g. extent --
                           1: < 5%; 2: 5-50%; 3: > 50%)

                    General Comments

[   ]
[   ]
[   ]
[   ]

Fig. 14.2
Sample data collection form for crop and land use studies. (Source: Curtis and Hooper, 1974.)

The range of data to be collected should also be related to the organisational structure and personnel resources. A summary table (Table 14.2) gives man-power requirements for ground data collection in respect of remote sensing studies in Gloucestershire, Somerset and Nottinghamshire. The prime objective of ground data collection is to provide a contemporaneous record of ground conditions at the time of imagery. In practice it is rarely possible to obtain detailed synchronous agricultural data for more than a small area or selected sample sites. In planning ground data collection, special attention should be given to the rate of change of the variables to be observed. These variables can be categorized as transient or non-transient. Data recording of transient features (e.g. crop stage, leaf cover) must be near synchronous. For example, data for spring barley in Nottinghamshire showed that mean per cent leaf cover increased from 18 to 40 in a period of 8–10 days in the first half of May. These changes are of sufficient magnitude to necessitate repetition of ground data collection since the proportion of bare soil exposed beneath a growing crop will have a major effect on image response. Quantitative observations of soil exposure by quadrating methods are time consuming. A single observer measuring crop cover using a 50 cm x 50 cm, 100 point, quadrat for 500 observations per field could cover approximately 25 fields per day. The most successful method of estimating leaf cover in a cereal crop such as barley may be to establish a relationship between crop stage and leaf cover for a sample of fields by quadrat measurements. Data for crop stage/leaf cover relationships are shown in Fig. 14.3, leaf cover in each field having been determined by 500 point observations.

Purdue University Laboratory for Agricultural Remote Sensing (LARS) record crop species, row width and direction, height of crop, percentage crop cover, and general crop condition. After the area has been overflown the LARS investigators also interview farmers to obtain information on planting technique, date of planting, date of harvest and the variety of species grown. The ground truth data is then key-punched on computer cards, processed and stored on magnetic tape. A computer programme then provides a print out on any particular field or field of interest, or for specific information for a particular group of fields (Fig. 14.4).

14.3 Crop and land use studies using satellite data

Crop surveys from multiband satellite photography using digital techniques have been widely investigated. The sequence of operations includes data preprocessing, training sample selection, and classification by statistical pattern recognition (Fig. 14.5). In the data preprocessing phase two operations are generally necessary. First, measurement of the film density at a large number of discrete points in the frame, and conversion of these measurements into digital form. Second, registration of the density measurements so that each of the multiband picture samples is stored on tape in geometric coincid-

Table 14.2

Assessments of manpower requirements for ground data collection. (Source: Curtis and Hooper, 1974)

Area observed (km²)	Total fields	Total observers	Date	Sampling method	Sampling density	Prior training	Progress km h⁻¹	Fields/ hour/ man
635	933	10 (Working in pairs)	June/July	Line traverse (road)	1 field in 4 along traverse	Agricultural officers familiar with area	8	22
70	341	1	May	Line traverse (road)	Continuous – all fields on traverse	Agricultural officers familiar with area	4.4	15
70	341	1	August	Line traverse (road)	Continous – all fields on traverse	Agricultural officers familiar with area	3.8	14

Sample study of the time allocation in ground truth data collection

Task	Percentage of total time	
	May (45 h)	August (46 h)
Ancillary data collection – soil samples and farming operations	12	5
Data collection and recognition on traverses	48	52
Data checking and office compilation of data interpretation of imagey	40	43

Relationship between growth stage and leaf cover in Spring Barley

Quadrat samples 6th 8th & 16th May 1973

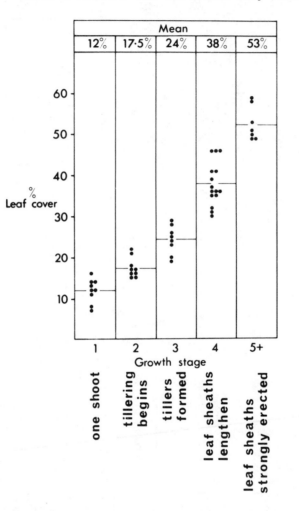

Fig. 14.3

Relationship between growth stage and leaf cover in Spring Barley. (Source: Curtis and Hooper, 1974.) p. 00)

SERIAL NUMBER......... 3666

FIELD DESIGNATION.... C-1-11

SIZE OF FIELD....   33 ACRES   SOIL TYPES..ELSTON L   ELSTON L   WEA SIL   ELSTON SIL

PRIMARY SPECIES.OATS   VARIETY. CLINTLAND

PLANTING TECHNIQUE...DRILLED

ROW DIRECTION...CIRCULAR   ROW WIDTH... 20 CM.

YIELD.....100 BUSHELS

PERTINENT DATES.		DENSITY FACTORS				TREATMENTS	
	DATE	CROP	HEIGHT CM	COVER PCT		DATE	TREATMENT
PLANTING MAR 15	MAY 6	OATS	005	00			
	JUN 28	OATS	080	80			
	JUL 25	STUBBLE	025	85			
	SEP 13	STUBBLE	030	99			

Fig. 14.4
Sample of computer print-out of ground truth data. (Source: LARS, 1968.)

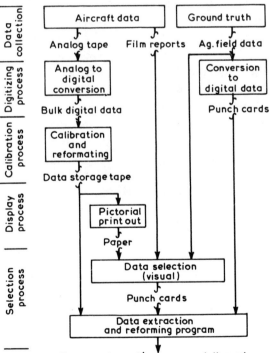

Fig. 14.5
LARS data flow system for crop studies.

ence. The training sample selection phase aims to determine the separable classes and subclasses in a given data set. Pattern analysis systems have been developed which allow several methods of class selection to be employed as appropriate. Statistics can be computed for up to 30 wavelength bands and printed out in the form of histograms, correlation matrices, and coincident one sigma spectral plots. One of these types of output can then be used to group areas having similar spectral responses. Another method of class separation consists of using clustering techniques to group image points such that the overall variance of the resultant sets is minimised.

The fields for which ground truth data is available are then identified by line and column coordinates so that they can be entered together with film density data into the computer. Histograms and statistics are then computed and printed for each field. An adequate statistical sample (> 30) for each crop type is necessary at this stage. Samples of data from each of the classes identified by the statistical process are then used for training the pattern classifier e.g. the histograms for barley are used as training sets in respect of the automatic classification of barley.

In the pattern recognition stage two methods can be used to classify the multiband imagery. One method classifies each image point into one of the defined classes. Alternatively an entire field can be classified as one decision. The latter technique, termed 'per field classification', has the advantage of speed. However it demands that the field coordinates must be known and fed into the classifier before any classification can be performed. Classification by the 'per point' method is time consuming since every resolution element in the image is classified separately. However it can classify any area without field boundaries being specified.

Following the pattern recognition phase, which produces automatic classification of the land use, it is necessary to evaluate the classification accuracy quantitatively. A large number of test fields are necessary for this purpose. These are located in the computer classification and the ground truth is compared with the computer result. An example of per field classification using the LARS system is shown in Table 14.3. The accuracies achieved in identifying land use over part of Imperial Valley, California varied considerably. The overall classification accuracy was 70.8 per cent, with peak accuracies of 100 per cent for bare soil and 82 per cent for barley. Identification of alfalfa was poor in all cases. It should be noted that the image data used was uncalibrated and there was considerable overlap of the spectral bands. Calibration of the data for atmospheric effects and better separation of the spectral bands may allow better crop separation accuracies.

Supervised classification techniques of the type described above have certain drawbacks. There is the very considerable difficulty that spectral signatures show high variability and supervised techniques generally demand that reference signatures be collected directly from a training area lying within the survey area or nearby. The unsupervized classification technique avoids this difficulty by not requiring reference signatures in the data processing phase. Unsupervised techniques will group the multi-

Table 14.3

Per field classification of test fields in the Dogwood Road area (Flightline 15A) utilizing data collected in the 0.51–0.79, 0.46, 0.68–0.89, 0.59–0.71 μm ranges.

Field no.	Ground truth classification	Computer classification	Field no.	Ground truth classification	Computer classification
2	Barley	Barley	267	Sugar beets	Sugar beets
6	Barley	Barley	289	Sugar beets	Sugar beets
12	Barley	Barley	290	Sugar beets	Sugar beets
17	Barley	Sugar beets	291	Sugar beets	Alfalfa
18	Barley	Barley	292	Sugar beets	Sugar beets
19	Barley	Barley	298	Sugar beets	Barley
20	Barley	Alfalfa	13	Alfalfa	Sugar beets
36	Barley	Barley	15	Alfalfa	Barley
47	Barley	Barley	14	Alfalfa	Barley
48	Barley	Barley	22	Alfalfa	Alfalfa
71	Barley	Barley	42	Alfalfa	Sugar beets
76	Barley	Barley	39	Alfalfa	Sugar beets
104	Barley	Barley	50	Alfalfa	Sugar beets
105	Barley	Sugar beets	58	Alfalfa	Sugar beets
119	Barley	Sugar beets	186	Alfalfa	Sugar beets
143	Barley	Bare soil	282	Alfalfa	Sugar beets
173	Barley	Barley	288	Alfalfa	Barley
198	Barley	Barley	1	Bare soil	Bare soil
199	Barley	Barley	5	Bare soil	Bare soil
213	Barley	Barley	7	Bare soil	Bare soil
227	Barley	Barley	8	Bare soil	Bare soil
240	Barley	Sugar beets	11	Bare soil	Bare soil
243	Barley	Barley	21	Bare soil	Bare soil
244	Barley	Barley	44	Bare soil	Bare soil
249	Barley	Barley	41	Bare soil	Bare soil
279	Barley	Barley	45	Bare soil	Bare soil
284	Barley	Barley	46	Bare soil	Bare soil
285	Barley	Barley	61	Bare soil	Bare soil
16	Sugar beets	Alfalfa	155	Bare soil	Bare soil
51	Sugar beets	Sugar beets	157	Bare soil	Salt flat
52	Sugar beets	Sugar beets	222	Bare soil	Bare soil
55	Sugar beets	Alfalfa	230	Bare soil	Salt flat
59	Sugar beets	Alfalfa	258	Bare soil	Bare soil
62	Sugar beets	Sugar beets	269	Bare soil	Sugar beets
66	Sugar beets	Sugar beets	98	Salt flat	Salt flat
77	Sugar beets	Sugar beets	101	Salt flat	Salt flat
154	Sugar beets	Sugar beets	106	Salt flat	Salt flat
163	Sugar beets	Sugar beets	125	Salt flat	Salt flat
164	Sugar beets	Sugar beets	126	Salt flat	Salt flat
245	Sugar beets	Barley	127	Salt flat	Salt flat
248	Sugar beets	Sugar beets	128	Salt flat	Salt flat
268	Sugar beets	Sugar beets	158	Salt flat	Salt flat
262	Sugar beets	Sugar beets	159	Salt flat	Salt flat
266	Sugar beets	Barley	221	Salt flat	Salt flat
			99	Water	Water

spectral data into a number of classes based on the same intrinsic similarity within each class. The meaning of each class in terms of land use category is then obtained after data processing by checking a small area belonging to each class. Preliminary tests of supervised and unsupervised classifications over areas in the United States suggest that the overall performance of each is comparable. If this conclusion is supported by future work it is likely that unsupervised classifications will be preferred because they work from the remote sensing data to the group instead of vice versa.

One of the most extensive crop studies so far made by remote sensing techniques, using supervised classification based on ground truth, is that of the Corn Blight Watch Experiment, 1971. The corn (maize) leaf blight is caused by the fungus *Helminthosporium maydis* and is widespread in maize growing in tropical areas of the world. Until 1969 corn leaf blight was a minor problem in the U.S.A. but in 1970 there was extensive corn blight as a result of the development a new race of the fungus. As a result yields

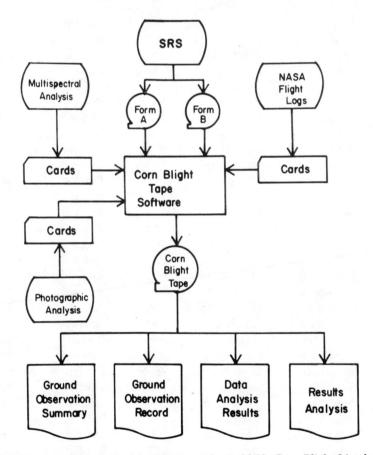

Fig. 14.6
Merging of ground data, flight logs and analysis results in 1971, Corn Blight Watch Experiment. (Source: MacDonald *et al.*, 1973.)

are thought to have dropped by about 700 million bushels in 1970.

In 1971 two aircraft collected colour infrared photography and 12-channel multi-spectral scanner imagery. The ground data, flight logs and the image analysis results were combined and analysed to record the results of the corn blight survey. (Fig. 14.6). The interpretation phase of this study required extensive correlations of field observation, photointerpretation and automatic analysis of MSS data. Five categories of corn blight severity were used in the field estimates as illustrated in Fig. 14.7a, and the results (Fig. 14.7b) showed that the best correlations were achieved when the blight symptoms were well established (stages 2,3,4). It also appears that the correlation between MSS analysis and field data became better at later dates in the study (Fig. 14.8). This important study concluded that neither manual photointerpretation of small scale photography nor machine made analysis of MSS data gave adequate detection of Corn Leaf Blight during early stages of infection. Analysis of the data did, however, permit the detection of outbreaks of moderate to severe infection levels.

Laboratory investigations by other workers have shown that maize leaves infected with corn blight possess similar reflectivities to healthy leaves in the visible range (0.4–0.75 μm). In the near infrared range (0.8–2.60 μm) healthy leaves show significantly higher reflectivities than *Helminthosporium maydis* infected leaves. Thus further research studies should investigate the infrared wavelengths for corn blight surveys.

Mapping of the spread of those levels of infection over the region could be carried out with relatively high accuracy.

There seems to be good evidence that land use, crop stress and crop disease can be mapped using satellite multispectral scanning data (Plate 14.3 see p.   ). However, ways must be found for reducing the amount of ground truth data required for initial selection of training samples. This may, again, point to increasing use of unsupervised classification. The problem remains, however, to devise classificatory methods which can be applied over a wide range of environmental conditions and over large geographic distances. There is also a need for further research to determine the effects of sun angle, observation angle and atmospheric effects on the spectral patterns of crops.

Considerable improvements in land use and crop condition, estimates may be possible when temporal variations in spectral patterns can be readily analysed. For example it can be shown that better interpretation can be achieved from 8 channels of data collected at three different times than 24 channels collected at one time. In order to make satisfactory analyses of such temporal data it will be necessary to achieve satisfactory registration of data so that the same areas can be compared.

14.4. Prehistoric land use

Aerial photographs have been widely used in studies of earlier civilizations and their

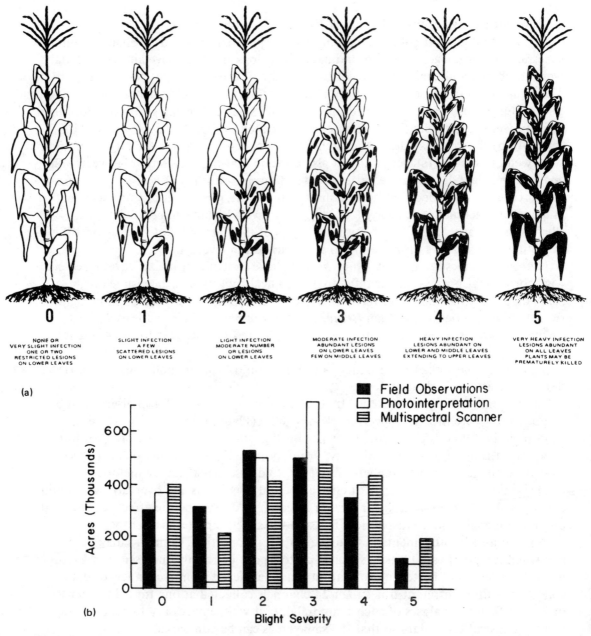

(a)

0	1	2	3	4	5
NONE OR VERY SLIGHT INFECTION ONE OR TWO RESTRICTED LESIONS ON LOWER LEAVES	SLIGHT INFECTION A FEW SCATTERED LESIONS ON LOWER LEAVES	LIGHT INFECTION MODERATE NUMBER OR LESIONS ON LOWER LEAVES	MODERATE INFECTION ABUNDANT LESIONS ON LOWER LEAVES FEW ON MIDDLE LEAVES	HEAVY INFECTION LESIONS ABUNDANT ON LOWER AND MIDDLE LEAVES EXTENDING TO UPPER LEAVES	VERY HEAVY INFECTION LESIONS ABUNDANT ON ALL LEAVES PLANTS MAY BE PREMATURELY KILLED

(b)

Fig. 14.7

(a) Scale for estimating southern corn leaf blight severity.

(b) Comparison of field observation, photointerpretation and machine analysis estimates of corn acreage in individual blight classes for the intensive study area, August 23 to September 5. (Source: MacDonald *et al.*, 1973.)

settlement patterns. In fact air photography is now an accredited, almost veteran, aspect of the archaeological method. Nevertheless, its potential and limitations are still not widely appreciated.

In practice oblique photographs are widely used to illustrate the features of archaelogical sites because they afford the advantages of a view comparable to that which could be obtained from a hill top in a hand picked position at close quarters. The vertical air photograph serves a different purpose in that it is the main tool for mapping features and detecting their planimetric relationships. When overlapping verticals are studied stereoscopically they offer many advantages in that the stereomodel enables the viewer to see the physiographic positions in which archaeological remains are found.

When air photographs are used to study former land use patterns there are two principal lines of evidence. First, there are vegetation markings which mainly occur within cropped land. Where buried features exist beneath the surface the soil may be shallow e.g. over a buried wall. In this case the rooting depth of crops will be restricted and at times of water deficiency (usually late summer) the crops may show yellowing of the leaves or stunted growth. Conversely the buried feature may be in the nature of a ditch. Under these circumstances the soil over the ditch may be more moisture retentive and higher in nutrients. In consequence the crop may be more luxuriant and greener along the ditch line. Such features are sometimes referred to as positive (stronger growth) or negative (weaker growth) crop marks. They are usually best seen when the crop is mature and when the climatic conditions have produced moisture stress in the soil (Plate 14.4).

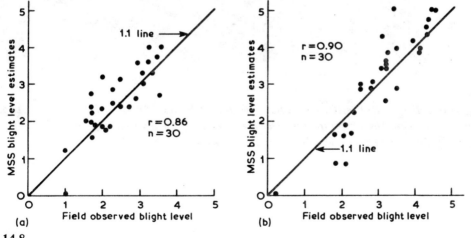

Fig. 14.8

Correlation of field observation and machine-assisted analysis of multispectral scanner data estimates of segment average blight severity levels, August 23 (left) and September 6 (right). (Source: MacDonald *et al.*, 1973.)

Plate 14.4
Crop marks revealing ancient settlement in fields west-north-west of Andover, Hampshire. Note also the pattern of wheel marks left by tractors near the gateway to the light-tones field on the right.
(Source: Cambridge University Collection, copyright reserved.)

Plate 14.5
A pattern of medieval fields in ridge and furrow near Husbands Bosworth, Leicestershire. Note how the low sun angle casts shadows which serve to emphasize small changes in surface relief. (Source: Cambridge University Collection; copyright reserved.)

The second type of evidence for prehistoric features consists of soil markings. These occur where the soil is bare or has only the thinnest covering of vegetation. The disturbed soil along field boundaries or foundations is different in structure, texture and colour from that of the undisturbed soil adjoining. These differences are often sufficient to induce different reflectances from the surface which can be detected by remote sensors. In England some of the greatest contrasts have been seen on chalk and limestone soils. Usually soil markings are most clearly detected after a period of exposure to wind and rain rather than immediately after ploughing.

The patterns of ancient field systems are often incorporated into present day field patterns. It is not always easy to distinguish old field boundaries from maps but aerial photography can often provide startling pictorial evidence for such features. For example the Roman field systems can often be observed in great detail. The Roman method of land partition was to allocate blocks of land which were usually in the form of a square (the centuria quadrata) of 710 m size. High altitude photographs show the grid pattern of centuriation well and in some cases lower altitude observations allow detection of earlier Greek systems to be identified. Similarly the patterns of ancient strip fields can be mapped from remote sensing data (Plate 14.5).

In this brief discussion of the use of remote sensing for the study of prehistoric land use attention has been focussed on the photographic sensors. There seems little doubt, however, that the application of infrared scanning systems could provide additional material of interest in such studies. These sensors are sensitive to changes in moisture status and could provide evidence of buried features.

References

Bell, T.S. (1974), 'Remote sensing for the identification of crops and diseases', in *Environmental Remote Sensing: applications and achievements* Barrett, E.C. and Curtis, L.F. (ed.), Edward Arnold, London, pp. 154-166.

Brenchley, G.H. (1964), 'Aerial photography for the study of potato blight', *World Review of Pest Control,* 3, 68.

Brenchley, G.H. (1968), 'Aerial photography for the study of plant diseases', *A. Review Phytopath.* 6, 1.

Brooner, W.G. and Simonett, D.S. (1971), 'Crop discrimination with color infrared photography: a study in Douglas County, Kansas', *Remote Sensing of Environment,* 2, 21.

Colwell, R.N. (1956), 'Determining the prevalence of certain cereal crop diseases by means of aerial photography', *Hilgardia,* 26, 223.

Curtis, L.F. and Hooper, A.J. (1974), 'Ground truth measurements in relation to aircraft and satellite studies of agricultural land use and land classification in Britain', *Proceedings Frascati Symposium on European Earth-Resources Satellite Experiments,* European Space Research Organisation, pp. 405-415.

Haralick, R.M. and Kelly, G.L. (1969), 'Pattern recognition with measurement space and spatial clustering for multiple images', *Proc. I.E.E.E.,* 57, 654.

Haralick, R.M. and Shanmugam, K.S. (1973), 'Combined Spectral and Spatial Processing of ERTS Imagery Data', *Remote Sensing of Environment,* 3, 3.

L.A.R.S. (1968), 'Remote multispectral sensing in agriculture', Laboratory for Agricultural Remote Sensing, *Purdue University, Research Bulletin,* 844.

Macdonald, R.B. *et al.* (1973), 'Results of the 1971 Corn Blight Watch Experiment', *Proceedings of the 8th International Symposium on Remote Sensing of Environment,* Ann Arbor, Michigan, 1, 157.

Meyer, M.P. and Calpouzos, L. (1968), 'Detection of crop diseases', *Photogrammetric Engineering,* 34, 554.

Nagy, G. and Shelton, G. and Tolaba, J. (1971), 'Procedural questions in signature analysis', *Proceedings of the 7th International Symposium on Remote Sensing of Environment,* Ann Arbor, Michigan.

Robbins, C.R. (1934), 'Northern Rhodesia: an experiment in the classification of land with the use of aerial photographs', *J. Ecology,* 22, 88.

Sisam, J.W.B. (1947), '*The Use of Aerial Survey in Forestry and Agriculture',* Imperial Forestry Bureau, Oxford.

Su, M.Y. and Cummings, R.E. (1972), 'An unsupervised classification technique for multispectral remote sensing data', *Proceedings of the 8th International Symposium Remote Sensing of Environment,* Ann Arbor, Michigan, pp. 861-879.

U.S.D.A. Soil Survey Staff, (1951), '*Soil Survey Manual',* United States Department of Agriculture, Handbook No. 18.

Forestry and ecological studies

15.1 Introduction

There is now a growing awareness that man's use of nature's processes and resources must be adjusted to the limitations and requirements which nature sets for us. Thus 20th century man has now become interested in the ecological changes which human settlements and industry have brought about in the local and the world environment. In particular he is now concerned with the conservation of species in danger of extinction and the maintenance of ecological balance in environments which have been altered to suit his economic needs. The important and fundamental fact remains that all animal life, including man, ultimately depends on the plant life which alone is able to synthesize elements into the form of food.

Wherever conditions are sufficiently favourable the climax vegetation cover consists of forest. In some areas, notably in savanna regions, forest clearance has led to an extension of grassland where forest formerly occurred. A summary of data for the Earth's cover of natural vegetation is shown in Table 15.1 from which it will be seen that approximately 42 per cent of the total land area is potentially forest land, 24 per cent potentially grassland and 34 per cent essentially desert.

Timber has been cut for a variety of purposes, mainly for fuel, construction materials and pulp. Many values are associated with forests besides the lumber and pulp that can be obtained from them. These include the maintenance of watersheds, oxygen production, their function as reservoirs for a variety of plant species, fish and wildlife, and the recreational and aesthetic pleasure they can provide.

The grasslands carry the vast numbers of livestock on which much of the world depends for protein-foods. Most of the critical problems in the management of grazing lands (range resource management) and husbandry of grazing animals occur in remote areas where least is known about native grazing land. In most cases what is needed is better information on (1) acreages of useable grazing land by ecologically appropriate classes, (2) the ecological characteristics of each kind of range (phytosociology, plant succession, present range condition, autecology of the important species) (3) the special management problems associated with different grassland ranges (4) indices of potential productivity of each kind of range. Also information is needed on the numbers and kinds of animals that make use of grassland resources. These include not only the farm animals but also the wild herbivores.

Table 15.1

Natural vegetation cover. (Source: Roberts and Colwell, 1954).

	Area in sq. miles	Percent of total land area
Forests		
Tropical rain forest	3 800 000	7.5
Temperature rain forest	550 000	0.9
Deciduous forest	6 500 000	12.0
Coniferous forest	7 600 000	15.0
Monsoon (dry) forest	2 000 000	3.8
Thorn forest	340 000	0.6
Broad sclerophyll forest	1 180 000	2.1
Total forest	21 970 000	42.0 (see note below)
Grasslands		
High grass savanna	2 800 000	5.3
Tall grass savanna	3 900 000	7.5
Tall grass	1 580 000	3.1
Short grass	1 200 000	2.4
Desert grass savanna	2 300 000	4.3
Mountain Grassland	790 000	1.4
Total grassland	12 570 000	24.0 (see note below)
Deserts		
Desert shrub and grass	10 600 000	21.0
Salt desert	30 000	—
Hot and dry deserts	2 400 000	4.7
Cold desert (tundra)	4 400 000	8.3
Total desert	17 430 000	34.0

Note. Some areas of potential grazing land occur within areas of
forest and desert. A more realistic figure for total land area
of potential grazing is 46 percent.

15.2 Remote sensing studies of forest conditions

In the field of forestry remote sensing studies by aerial photography have been used
for several decades. Air photo interpretation has been used for classification of forest
stands and types, survey of mortality and depletion, planning of reforestation, inventory
of timber and other forest products and assessment of property taxes.

As well as using photographic tone, texture and colour to identify different tree
stands foresters can make precise measurements of tree height, crown diameter or stand

Table 15.2

Example of an aerial stand volume table for Rocky Mountain conifer species. Gross cubic footvolume/acre by average stand height, average crown diameter and crown cover. (Volumes in tens of cubic feet.) (Source: *Manual of Photographic Interpretation,* 1960).

Average stand height (feet)	Crown cover (per cent)									
	5	15	25	35	45	55	65	75	85	95
18– to 22–foot crown diameter										
30		30	50	70	80	90	105	120	140	160
35		40	65	80	90	105	120	140	160	180
40	10	50	75	90	105	120	140	160	180	200
45	20	60	85	105	120	140	160	180	200	220
50	35	75	100	120	140	160	180	205	225	245
55	45	85	115	140	165	190	210	230	250	270
60	65	115	150	180	205	225	245	265	285	305
65	85	150	190	220	240	260	275	290	310	330
70	110	190	230	255	275	290	310	330	350	370
75	160	240	270	295	310	325	345	365	385	405
80	195	275	305	330	350	370	390	410	430	450
85	250	330	360	385	405	425	440	455	475	495
90	290	370	405	430	450	470	485	505	530	550
95	340	420	450	480	500	520	540	560	580	600
100	390	470	500	530	550	570	590	610	630	650
105	460	540	580	600	620	640	655	670	685	705
110	520	600	640	660	680	700	720	740	760	780

density. Tree height is closely correlated with tree volume and stand volume and can be measured on photographs in a number of ways. Measurement of shadows, measurement of parallax and measurement of relief displacement on single large-scale vertical or oblique photographs are some of the methods used. Tree shadow measurements are normally made with a micrometer scale consisting of a finely graduated series of short lines, one of which is matched with each shadow visible on the air photograph. Measurement of crown diameter can be made by either micrometer or dot type crown wedges. The micrometer wedge used normally consists of two converging lines calibrated to read intervening distance to the nearest throusandth of an inch. The dot type wedge usually consists of a series of dots differing in diameter by 0.0025 in. It is laid alongside the image and moved until the dot which just matches the size of the crown is identified. Alternatively the dot images can be moved over the crown until the appropriate size which covers the crown is found. The accuracy of crown measurement is largely dependent on the scale of the photographs. The error may be 3–4 feet with either kind of measuring device on

Table 15.2 *continued*

Average stand height (feet)	Crown cover (per cent)									
	5	15	25	35	45	55	65	75	85	95

23+ foot crown diameter

30	10	50	75	90	105	125	145	165	190	215
35	20	65	85	105	125	145	165	185	210	235
40	30	75	100	120	140	160	180	205	230	255
45	40	85	110	135	160	180	200	225	250	275
50	55	100	130	160	185	205	225	245	265	285
55	70	120	160	190	210	230	250	270	290	310
60	85	155	195	225	245	265	285	305	325	345
65	120	190	230	255	275	290	310	330	350	370
70	155	225	265	290	310	330	345	365	385	405
75	205	275	305	330	350	365	380	400	420	440
80	245	315	345	370	390	405	420	440	460	480
85	290	360	395	420	440	455	470	490	510	530
90	340	410	440	465	485	505	525	545	565	585
95	385	455	485	515	535	555	575	595	615	635
100	440	510	540	570	590	610	625	640	660	680
105	510	580	610	640	655	670	685	700	715	730
110	660	640	670	690	705	720	735	750	770	785
115	675	690	720	740	760	775	790	805	820	835

photographs of 1: 12 000 scale.

Transparent dot templets are probably the most widely used area measuring instruments in forest inventories made from aerial photographs. The density of dots in a templet varies from 1−65 per square inch depending on the intensity of the survey and the scale of the maps on photographs used. The ratio of dots in a given class to dots in the entire tract gives the proportion of the tract occupied by that class.

Crown coverage or crown closure, expressed as a per cent of the area covered by trees on the image is often considered a better estimate of stand density than crown coun A transparent overlay carrying a dot grid, fine enough to place 25−60 dots in the forest stand is placed over the image. The number of dots falling on tree crowns is then compared with the total number of dots in the stand area.

These elementary devices for the measurement of size and area can now be replaced (at a cost) by automated scanning equipment. If very large areas are to be studied such equipment is essential for data to be available within a reasonable period of time.

Once average height, crown diameter and crown cover are know it is possible to estimate the stand volume of timber from tables e.g. Table 15.2. The accuracy of the estimate will depend on the accuracy of the measurements made from the image and on the goodness of the correlations of tree height, diameter, and crown cover with volume of timber as shown in tables.

As in agricultural and soil studies the development of colour and infrared colour proved to be very useful in forest studies. In the infrared wavebands there is strong reflectance from deciduous trees but lower reflectance from coniferous trees. Thus both infrared black and white film and infrared colour film can be used to make broad distinctions between deciduous and coniferous trees (Plates 15.1). Examples

Plate 15.1
Multispectral photograph of farmland near Thetford, Norfolk. Note the differentiation between deciduous (light tones) and coniferous (dark tones) trees in the Band 4 image. (Courtesy, Hunting Surveys Ltd., Boreham Wool Herts and Natural Environment Research Council.)

of the use of colour and infrared colour in studies of forest type and inset damage have been given by many authorities. Dead or dying trees usually image in blue-grey instead of red colours (Plate 15.2 see colour plates).

Leaves vary from those with an extremely glossy cuticle where reflectance is largely specular, to those with essentially matte surfaces where reflectance is almost entirely due to scattering. Most field situations, however, result in a mixture of specular and diffuse reflectance. The leaf constituents are usually relatively transparent and, with the exception of the pigments found in the chloroplasts, do not absorb significant amounts of energy in the visible and near infrared wavelengths (0.4–0.9 μm). The typical leaf anatomy (Fig. 15.1), provides many air-water interfaces between the cell walls and the intercellular spaces. The index of refraction from 1.33 for water in the cell walls to 1.00 for air in spaces leads to effective internal reflections.

Although most leaf components are relatively transparent absorption takes place at those wavelengths required for photosynthesis. The predominant leaf pigments, chlorophyll a, chlorophyll b, carotene and xanthophyll all absorb in the vicinity of 0.445 μm i.e. the blue part of the spectrum. In addition chlorophyll absorbs in the vicinity of 0.645 μm i.e. the red part of the spectrum (Fig. 15.2). A small part of the energy absorbed by chlorophyll undergoes wavelength change and is emitted as fluorescent energy. Another part is converted photochemically into stored energy in the form of organic compounds.

Some representative spectral reflectance curves for different tree species are given in Fig. 15.3. It is clear that the multispectral approach towards recognition of tree categories is worth pursuing. Hoever, some difficulties attached to classification of vegetation using existing ERTS-1 multispectral data have been observed. Although computer analysis can provide good recognition of vegetation classes that have mature and uniform canopies at the time of data collection it has proved difficult to classify senescent vegetation. In other words where there is a non-uniform distribution of dead and dying vegetation along with patches of more healthy vegetation the classification process becomes difficult. In these circumstances the accuracy of classification from remote sensing data depends on the stage of growth and optimum times for collecting data will vary from one crop to the next.

15.3 Phenological studies

In view of the importance of the stage of growth for the interpretation of spectral responses it has become necessary to consider the detection of phenological events for specific forest and crop types. The term phenology has been applied to that branch of science which studies periodic phenomena in the plant and animal world insofar as they depend upon the climate of any locality. A phenology satellite experiment using ERTS-1 data has been described by NASA investigators. In their study two phenological sequences were observed:

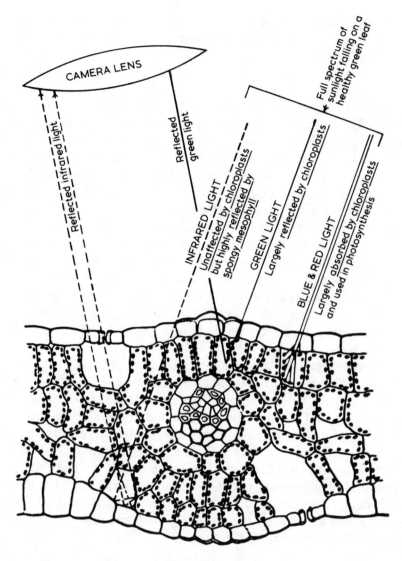

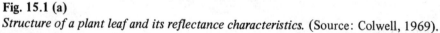

Fig. 15.1 (a)
Structure of a plant leaf and its reflectance characteristics. (Source: Colwell, 1969).

(a) The Green Wave. A record of the geographical progression with time of foliage development in plants over wide areas.

(b) The Brown Wave. A record of the geographical progression with time of vegetation senescence (maturation of crops, leaf colouration and leaf abscission). This progression plays the analogous role in the autumn as the Green Wave.

Results to date from the Phenology Satellite Experiment show that is is possible to

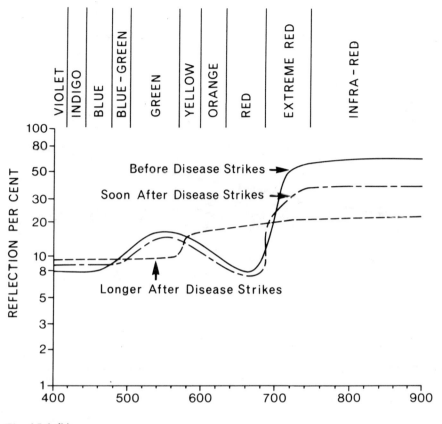

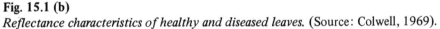

Fig. 15.1 (b)
Reflectance characteristics of healthy and diseased leaves. (Source: Colwell, 1969).

develop phenoclimatic models of the type shown in Fig. 15.4. For countries with a highly developed agriculture such information would be useful in determining crop status, yield prediction and management planning. Preliminary tests also indicate that bands 5 and 7 of Landsat imagery can be used to map events such as crop harvest and leaf fall for specific areas and possibly for entire regions.

In grassland areas existing Landsat imagery does not provide sufficient detail in mapping for evaluating management problems or making management decisions with sufficient confidence. The most important use of Landsat imagery for grassland ranges is that of monitoring changes in forage plant conditions and development (i.e. Green Wave/Brown Wave effects). It has been found that the approximate time of germination and the area where germination has occurred can be determined from ERTS MSS data. Also it is possible to determine areas and dates when plant growth ceases due to drought or soil moisture depletion.

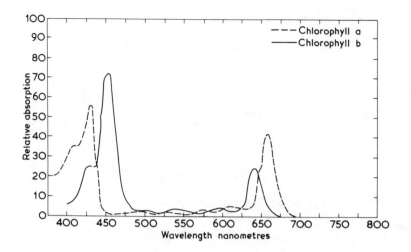

Fig. 15.2
Curves showing relative absorption of chlorophyll a and chlorophyll b as a function of wavelength. The combined absorption would be a summation of the two curves at each wavelength. (Source: Colwell, 1969.)

MAPLE, GREEN

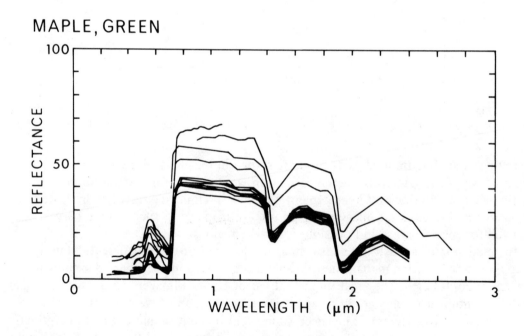

Fig. 15.3
Spectral reflectance curves of selected trees. (Source: Leeman *et al.*, 1971.)

OAK, GREEN

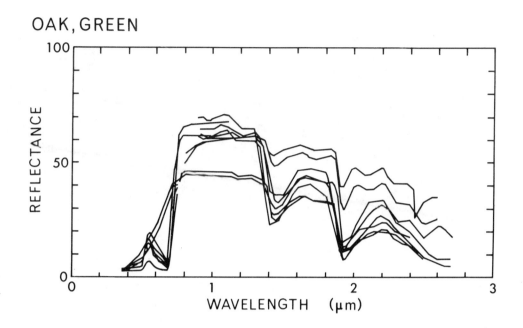

SYCAMORE, LIVE

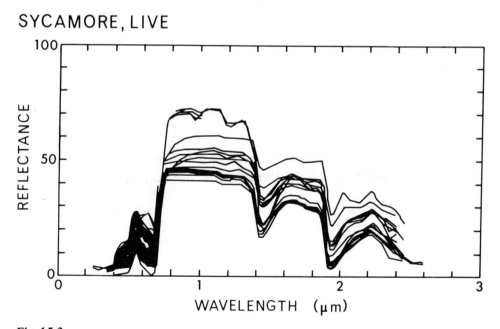

Fig. 15.3
Spectral Reflectance Curves. (Source: Leeman *et al.*, 1971.)

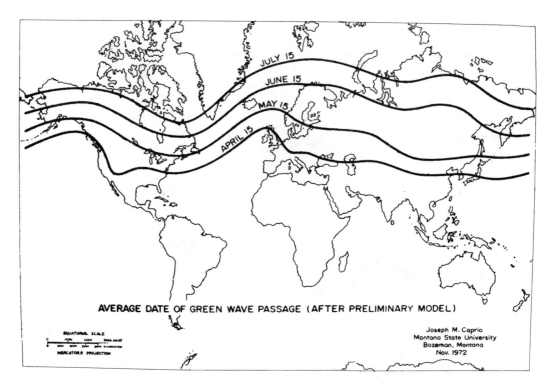

Fig. 15.4
Average date of the Green Wave (greening of grass) passage in the northern hemisphere. (Source: Dethier *et al.*, 1973.)

15.4 Wildlife studies

The most important uses of remote sensing studies in wildlife management are for making censuses of animal populations and mapping and evaluating the vegetation in the area occupied by the game. Inventories of wildlife populations are necessary for the planning of management programmes and formulation of fishing and hunting regulations. When surveys are made by ground methods animals may be counted twice or perhaps not at all. Aerial surveys minimize such errors by making it possible to cover an area quickly and completely. Therefore aerial surveys by direct observation were made as early as the 1930's in the United States. The use of aerial photographs developed somewhat later but gradually became important. The animals counted by air photo methods include antelope, deer, elk, barren-ground caribou, moose, muskoxen, waterfowl, seals, and in certain circumstances fish. It is interesting to note that one of the main difficulties facing wildlife researchers was the task of counting thousands of animals. This was particularly the case with photographs of waterfowl where some 14 000 geese may be recorded on a single photograph. Various attempts were made to develop counting methods, e.g. the scanning device of Kalmbach, 1949, which allowed successive slits to be examined and the animals were counted by a binocular microscope. Another method consisted of placing transparent overlays over the print and pricking the image of each animal counted.

It is clear that the development of automatic scanning densitometers allows such tedious (and sometimes inaccurate) counting to be superseded. Automatic counting techniques are now readily available for such work. However before such counting can take place it will be necessary to identify the characteristic spectral responses for different animals. The application of colour and black and white photography to the inventory of livestock has been studied by workers in the U.S.A. A comparison between image counts (I.C.) and ground enumeration (G.E.) for an area of about 1,000 square miles in Sacramento Valley, California is shown in Table 15.3.

Table 15.3

Comparison of image counts with ground enumeration of livestock numbers for area sampling units – cultivated stratum. (Source: Roberts and Colwell, 1968).

Sampling unit number	Livestock species							
	Cattle		Sheep		Other		All species	
	I.C.	G.E.	I.C.	G.E.	I.C.	G.E.	I.C.	G.E.
1	45	47	0	0	0	0	45	47
2	0	0	0	0	0	1	0	1
3	0	0	0	0	0	0	0	0
4	22	25	184	180	0	3	206	208
5	135	182	0	1	8	1	143	184
6	202	255	0	0	15	0	217	255
7	0	0	0	0	0	0	0	0
8	0	0	0	0	0	0	0	0
9	0	0	0	0	0	0	0	0
10	2	0	0	0	0	4	2	4
11	0	0	0	0	0	0	0	0
12	58	61	314	608	6*	0	378	669
13	0	0	0	0	0	0	0	0
14	0	0	1373	1000	0	0	1373	1000
15	0	0	0	0	0	0	0	0
16	41	43	0	0	5*	1	46	44
Totals	505	613	1871	1789	34	10	2410	2412

*Cattle incorrectly identified as horses

Infrared scanning techniques are of interest in livestock studies in that they offer a potential means for counting and studying the distribution of animals at night. The body temperatures of animals normally contrast with the surface ground temperature (Plate 4.4b). It is clear that thermal scanning used alongside aerial photographic methods could also be a means of estimating mortality rates. This idea has been proposed for counting dead seals after culling operations.

Radar studies of migrating swarms of locusts began when they were accidentally seen on radar over the Persian Gulf. It is well known that long distance migrations are made by swarms of sexually immature locusts. Apart from these swarms, however, locusts can also exist in the soliary phase and the solitaries fly by night and rest by day. From field studies using radar by night in the Niger Republic of Africa it has been shown that solitary locusts are able to orient themselves down-wind in contrast to the essentially random orientation within a swarm.

Although extensive survey by radar over large areas is not economic at present locust research workers have suggested that radars placed at strategic points — for example on frequently used migration tracks or regions where control operations are taking place — could enable population estimates to be made. In this way protective measures could be planned.

Studies of locusts have also shown that for substantial changes in numbers to occur (i.e. plague development) a sequence of favourable rains must fall in the breeding areas that can be reached by the migration of the apparently less mobile solitary locusts. It is, therefore, desirable to obtain records of wet ground or areas where green vegetation indicates recent rains in desert areas. An experiment has been made to compare Landsat observations of changing albedo and ground colour with changes in soil moisture and vegetation greeness in south western Saudi Arabia. Results have not yet been assessed but it is hoped that edges of wet or green areas will be detected. If it is found feasible to identify these wet areas the potential breeding sites for locusts can be surveyed.

The application of remote sensing techniques can do much to monitor the conditions within the wildlife habitat. It is possibly in this area that most assistance will be forthcoming for those concerned with wildlife management. The problem facing many conservationists is that measurements of changes in an ecosystem are time consuming and often only occur at a few points. Remote sensing methods can provide considerable detail over wide areas and can, therefore, aid in decisions concerning wildlife management.

References

American Society of Photogrammetry, (1960), *Manual of Photographic* Interpretation, Washington.

Bradford, J. (1957), *Ancient Landscapes: Studies in Field Archaeology,* Bell, London.

Carneggie, D.M., Poulton, C.E. and Roberts, E.H. (1967), *The Evaluation of Rangeland Resources by Means of Multispectral Imagery,* School of Forestry and Conservation, University of California, p. 76.

Carneggie, D.M. and De Gloria, S.D. (1973), Monitoring California's forage resource using ERTS-1 and supporting aircraft data, *Symposium on significant results from ERTS-1,* NASA, pp. 91-95.

Colwell, R.N. *et al.* (1963), Basic matter and energy relationships involved in Remote Reconnaissance, *Photogrammetric Engineering,* **29**, 761.

Colwell, R.N. *et al.* (1969), *Analysis of Remote Sensing Data for Evaluating Forest and Range Resources,* School of Forestry and Conservation, University of California, p. 207.

Dethier, B.E., Ashley, M.D., Blair, B. and Hopp, R.J. (1973), 'Phenology satellite experiment', *Symposium on significant results from ERTS-1,* NASA, pp. 157-165.

Gates, D.M. (1965), 'Heat transfer in plants', *Scient. Am.,* **213**, 76.

Heller, R.C. (1968), 'Large scale color photography samples forest insect damage', in *Manual of Color Aerial Photography,* Smith, J. (ed.), p. 394.

Howard, J.A. (1970), *Aerial Photo-Ecology,* Faber, London.

Jensen, C.E. (1948), 'Dot-type scale for measuring tree crown diameters on aerial photographs', U.S. Forest Service, Central States Forest Experiment Station, Note No. 48.

Kalmbach, E.R. (1949), 'A scanning device useful in wildlife work', *J. Wildlife Management,* **13**, 226.

Leedy, D.L. (1953), 'Aerial photo use and interpretation in the fields of wildlife and recreation', *Photogrammetric Engineering,* **19**, 127.

Leedy, D.L. (1968), 'The inventorying of wildlife', in *Manual of Color Aerial Photography,* Smith, J. (ed.), p. 422.

Myers, V.I. and Allen, W.A. (1968), 'Electro-optical remote sensing methods as non-destructive testing and measuring techniques in agriculture', *Applied Optics,* **7**, 1819.

O'Neill, H.T. (1953), 'Keys for interpreting vegetation from air photographs', *Photogrammetric Engineering,* **19**, 422.

Pedgley, D.E. and Symmons, P.M. (1968), 'Weather and the locust upsurge', *Weather,* **23**, 484.

Pedgley, D.E. (1974), 'Use of satellites and radar in locust control' in *Environmental Remote Sensing: applications and achievements,* Barrett, E.C. and Curtis, L.F. (eds.), Edward Arnold, London, pp. 143-152.

Roberts, H. and Colewell, R.N. (1968), 'The Application of Remote Sensing to the Inventory of Livestock and Identification of Crops', School of Forestry and Conservation, University of California, p. 20.

Roffey, J. (1969), *Radar studies on the Desert Locust,* Anti-Locust Research Centre Occasional Report, 17/69.

Rogers, E.J. (1949), 'Estimating tree heights from shadows on vertical aerial photographs', *J. Forestry,* **47**, 182.

Safir, G.R. and Myers, W.L. (1973), 'Application of ERTS-1 Data to analysis of agricultural crops and forests in Michigan', *Symposium significant results from ERTS-1,* NASA, pp. 173-180.

Schaefer, G.W. (1972), 'Radar detection of individual locusts and swarms', *Proceedings of the International Study Conference on the Current and Future Problems of Aridology,* London, 1970, pp. 379-380.

Seeley, H.E. (1948), *The Pole Scale,* Dominion Forest Service (Canada) Forest Air Survey Leaflet No. 1.

Shantz, H.L. (1954), 'The place of grasslands in the Earth's cover of vegetation', *Ecology,* **35**, 143.

Spurr, S.H. (1960), *Photogrammetry and Photointerpretation,* Ronald Press, New York.

Thorley, G.A. (1968), 'Some uses of color aerial photography in Forestry' in *Manual of Color Aerial Photography,* Smith, J. (ed.), p. 393.

Wilson, R.C. (1948), 'Photo interpretation aids for timber surveys', *J. Forestry,* **46**, 41.

Worley, D.P. and Meyer, H.A. (1955), 'Measurement of Crown Diameter and Crown Cover and their accuracy on 1: 12,000 scale photographs', *Photogrammetric Engineering,* **21**, 372.

 **Urban studies**

16.1 General considerations

Urban centres have been described as complex associations of population concentrations, intensive activities, and diverse lifestyles. One of the principal tasks that must be undertaken in order that cities might be made more attractive and convenient places in which to live is the compilation of an adequate fund of information on cityscapes as they appear today. Often town planners are called upon to suggest improvements to an urban morphology which is inadequately summarized by existing information. The acquisition and analysis of urban data other than those provided by standard census returns can be both costly and time-consuming. More often than not the data which are available relate very specifically to a particular aspect of the urban scene or urban life, and cannot be used for other studies in the absence of more suitable information. Remote sensing affords a rapid means whereby information of a less selective kind can be obtained. Here the types of systems used and the resolutions of the resulting data become the selective filters, rather than the wording of questionnaires or, perhaps, the non-cooperation of sectors of the urban community in answering them.

For many years aerial surveys of cities have been carried out photographically, at first in black-and-white, and more recently in some cases in colour. Unfortunately such surveys have been restricted to certain areas of the world, notably the more advanced industrial nations, and have been both piecemeal and infrequently repeated even there. We may say that the urban centres in which planning and renewal might have the greatest beneficial effects are mostly those which have been surveyed least frequently or even not at all.

Two nonphotographic remote sensor systems have been used with increasing frequency in urban studies in recent years, namely thermal infrared systems and radar. Many land use and transportation features can be distinguished in infrared imagery. Radar is particularly valuable when bad weather would otherwise have led to the postponement of a photographic mission. Some areas of the world are notoriously cloudy. Here the opportunities in the course of the year for photographic or even infrared surveys may be very few, but radar may still supply useful data. Radar may be the only effective means whereby rapid assessments can be made of sudden changes in urban morphology, for example patterns of destruction caused by tornadoes, earthquakes, or enemy hostilities.

We may subdivide the information contents of remote sensing data of urban areas into classes of features which are directly observable, and those which are only indirectly observable or 'inferential'. An example of the first class is the total number of buildings of specified type in a selected area. An example of the second is the social class of a neighbourhood. In this case indirect evidence is used to infer a phenomenon associated with, or indicated by, the visible structures and paraphernalia.

In many instances urban information from remote sensing sources can be of little practical help unless its resolution, or detail content, is sufficiently high. In the past the resolutions of many radar and infrared systems flown on aircraft and spacecraft have been too coarse to satisfy the needs of urban planners and other council authorities. For this reason their use has been relatively restricted. The resolution threshold of satellites has been particularly disadvantageous, and even Landsat data have been of only marginal interest to students of townscapes or city regions. Although many attractive maps have been prepared from Landsat data these have revealed few new features, or relationships not previously known. Certain military satellites have yielded data of restricted spatial coverage with resolutions of metres, rather than tens of metres, but their information has not been released for general use amongst the scientific and/or civic communities. Hopefully, some of the studies which have been based upon them will eventually be released from the secret list to point the way to a much more widespread and useful programme of urban remote sensing from satellites, with all the advantages of repetitiveness and cheapness per unit operation that such a programme would possess.

16.2 Urban structure and morphology

16.2.1 Land use

Land use maps may be viewed as spatial inventories of how land is being utilized. Such maps are, perhaps, those consulted most frequently by urban planners. Remote sensing affords the means whereby both rapid and repetitive surveys can be made. We have remarked upon scale as a restricting aspect of imagery from the Lansat satellites. However, at times the resolution achieved by a specific sensor/platform/processing arrangement may be beneficial, not problematical, for mapping features of selected sizes. For example, if urban land use information is required only on a grid square or block basis, then a relatively low-resolution system will provide adequate information and be advantageous in eliminating the finer details of urban morphology which may be regarded as a form of picture 'noise'. Aerial photography at a scale as small as 1: 100 000 may suffice. On the other hand, if details of land use at the plot or parcel level are sought, then much larger scale photography is necessary, supplemented by additional sources of information. Table 16.1 summarizes some user requirements for land use data generally, in terms of areal unit sizes and the degrees of detail associated with land-use classification systems.

Table 16.1
Some user requirements for land-use data: user type versus area size and land use classification. (Source: Horton, 1974.)

User type	Functional type	Areal unit	L.U. classification*
National	Community development	Cities over 2 500 pop. Counties	one-digit
	Economic development	Cities over 2 500 pop. Counties	one-digit
	Human resources	Cities over 2 500 pop. Counties	one-digit
	Natural resources	Counties	one-digit urban two-digit nonurban
State	Community development	Townships	two-digit urban one-digit nonurban
	Economic development	Cities over 100 pop.	two-digit
	Human resources	Census tracts for cities over 25 000 pop.	two-digit urban one-digit nonurban
	Natural resources	40-acre parcels (1/16 section)	one-digit urban two-digit nonurban
Local	Community development	City block	four-digit urban two-digit rural
	Economic development	40-acre parcel (1/16 section)	four-digit
	Human resources	City block	four-digit residential two-digit other
	Natural resources	40-acre parcel (1/16 section)	two-digit urban four-digit non-residential

* One-digit classification divides land use into general categories, such as residential, industrial, commercial, agriculture, etc.; two-digit classification of land use breaks each general category into subdivisions such as single-family residence, two-family structures, multifamily structures, etc; four-digit classification provides an extremely detailed breakdown such as distinguishing among retail uses.

The most ambitious single project to evaluate the usefulness of photography from high-altitude aircraft and Earth-orbiting satellites for urban land use change detection is the Census Cities Project. This has been organized by NASA in conjunction with the Geographic Applications Program (GAP) of the U.S. Geological Service (USGS). Originally twenty-six cities were named as test cases, and the U.S. Air Force Weather Service and NASA's Manned Spacecraft Center acquired multispectral, high-altitude photography from twenty of them. The year chosen for this survey was 1970, the time of the ten-year U.S. Census. The basic remote sensing data were colour infrared

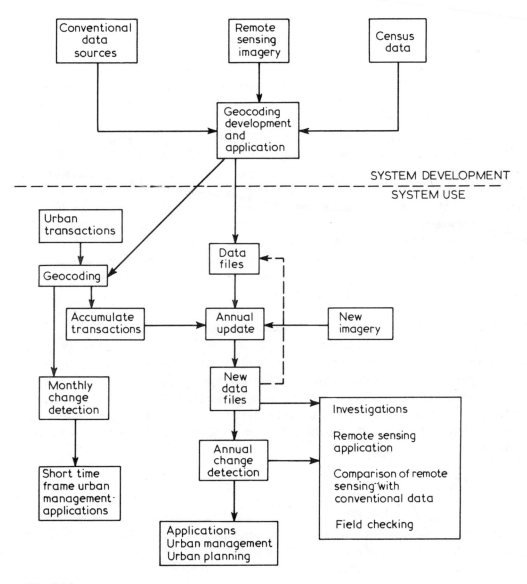

Fig. 16.1
A flow diagram for an operational urban change detection system. (Source: Horton, 1974)

photographs. This film type is especially valuable for use over towns on account of its high haze-penetration capability. A further useful characteristic of this type of film is that it distinguishes clearly between vegetation (reddish in colour) and cultural features (which have a blue appearance on the film). This makes it especially useful where cities have a strong urban structure/woodland mix. The ultimate goal of the Census Cities

Project is the production of an Atlas of Urban and Regional Change. This will accommodate many types of data presentation, including photo-mosaics, conventional maps, computer-printed maps, tabulated data and text. In a pilot study for the city of Boston, a 24-category land use classification was employed. The minimum cell size was about 10 acres. The land use data were computer-processed to make them compatible with the 1970 census data. Consequently they can be retrieved, either singly, or in selected combinations, by the census tracts.

At a smaller scale, Landsat imagery (Plate 16.1 see colour plates) have been processed to give land use maps of urban areas at scales from 1: 250 000 upwards. Such maps can be compiled very rapidly. For example, an 11-category map has been prepared from a single Landsat multispectral scanner image in only eight man days. Current research is investigating the best channel, or combinations of channels, in the MSS for such applications.

Fig. 16.1 represents a suggested system for the detection of change in urban land use patterns. Remote sensing imagery is vital to such a scheme, for this comprises the 'status report' information involved in the 'update' section of the scheme in operational use. Clearly snapshot data from high-altitude platforms should increase the dynamic element in studies of urban land use and morphology.

16.2.2 Population, housing quality, and family income

We may conveniently treat this triumvirate together on account of the close links which exist between them. In the previous section mention was made of census figures from conventional, not remote sensing, sources. Given adequate remote sensing data, and sufficient ground truth from sample areas, it should be possible to assess population densities and totals indirectly from photographic or multispectral imagery. One model to estimate important population parameters for small areas using variables derived from remote sensing imagery was applied to Washington, D.C. in 1970. The general model was formulated as:

$$Y = f(x_1, x_2, ..., x_n) \tag{16.1}$$

where Y represents housing unit counts or population (1970 tract statistics) and $x_1 ... x_n$ are imagery-derived variables such as the number of single-family structures, the number of multiple-family structures, distance from the central business district etc. Relationships between x and Y were developed by multiple regression analysis.

In practice, residential land use was identified on 1: 50 000 scale aerial photographs, and the residential land area was computed, A block-by-block count was then conducted to establish the number of dwelling units per structure according to a four-category classification (single-family housing units, 2–5, 6–14, and more than 15 housing units). Multiple regression coefficients of 0.65 and 0.54 were obtained for central city and surburban tracts respectively. These are encouraging, but point to the need for a number of improvements before such a scheme becomes operational. These include the provision of remote sensing imagery of a type in which tree cover poses less of a problem for the

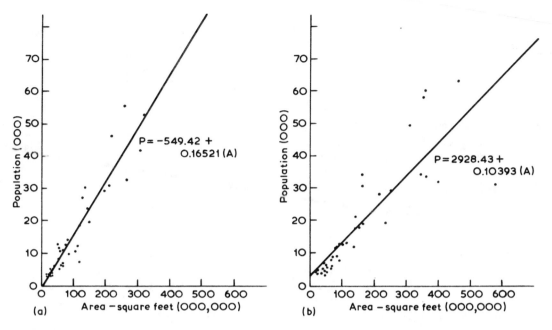

Fig. 16.2
Observed relationships between population and area of towns in the Tennessee River Valley, in (a) 1953 and (b) 1963. (Source: Holz *et al.*, 1969)

location and classification of housing units, plus a greater range of ground truth data for the formulation of the model and the checking of its output.

On a broader scale it has been proposed that remote sensing techniques could be used to estimate the total population of smaller towns over wide areas in intercensal periods. One test area for this proposal was the Tennessee River valley region for which aerial photographs were available for 1953 and 1963. Four hypotheses were treated. These were as follows:

(a) The population of an urban area is positively related to the number of links it has with other urban areas.

(b) The population of an urban area is positively related to the population of the nearest larger urban area.

(c) The population of an urban area is inversely related to the distance to the nearest urban area.

(d) The population of an urban area is proportional to the observable area of occupied space of such a population.

These hypotheses were tested using stepwise linear regression. It emerged that, except for urban area, the order of the independent variables differed from time to time, the values of the various coefficients differed significantly, and for one variable in particular (the distance to the nearest larger urban area) reversals of the direction or sign of the

Table 16.2

A comparison of housing areas in Austin, Texas, based on selected environmental criteria amenable to identification on remote sensing imagery. (Source: Davis *et al.,* 1973).

Criteria	Low-income areas	Middle-income areas
House size, (sq. ft)	380 – 1220 (average, 731)	1110 – 1560 (average, 1305)
Placement of house and lot (distance from street in feet)	12 – 42 (average, 27)	34 – 45 (average, 40)
Potential landholding per housing unit (sq. ft)	5670	7500
Building density (%)	14.2	17.7
Average lot size and frontage (sq. ft)	4337	7376
Image, pattern, and texture	Narrow frontage Lack of uniformity Irregular	Wide frontage Uniform Regular
Houses with driveway (%)	8.3	97.0
Houses with garage (%)	3.0	97.0
Number of visible autos per house	0.20	0.76
Unpaved street (%)	65	0
Street width (ft)	12 – 24 (average, 18)	24 – 31 (average, 28)
Quality of curbing	Generally lacking	Intact
Quality of vegetation	Not as vigorous	Cultivated, vigorous
Housekeeping	Presence of debris	Lack of debris
House orientation to street	Short side	Long side
City block pattern	Irregular, dead end twisting streets	Regular
Vacant lots per city block	0.6	0
Proximity to manufacturing and retail activity	Close to manufacturing Remote from shopping outlets	Remote from manufacturing Close to retail outlets

coefficient occurred. However, notwithstanding these complications, the multiple correlation coefficients between actual and estimated population for 40 selected central places were 0.95 for 1953 and 0.88 for 1963. These indicated that over 91 per cent, and 77 per cent respectively of the variations in population of the urban areas is explained on the bases of the selected independent variables. Because of the high correlation between urban area and population (see Fig. 16.2) the inclusion of the remaining independent variables adds little to the explanation of the variation that remained.

We may conclude that remote sensing, even at a coarse scale, seems destined to provide valuable information concerning population levels and changes, in most areas of the world. The application of aerial photography to dwelling-unit and population estimation would appear to have a particular potential in developing countries where rates of urban and population growth are especially great. Demographic data in such countries are frequently less than adequate for planning purposes, since national censuses are difficult to undertake and often yield inaccurate results.

A partly-related question of considerable significance to government and municipal authorities is that of housing quality. For example concerted attacks on the problems of urban poverty neighbourhoods cannot be planned — still less carried out — until such areas have been defined, and located on the ground. Conventional methods of amassing data on housing quality distributions are extremely time-consuming. Once again remote sensing techniques may provide new and better data than surveys on the ground, while providing such data more quickly and frequently.

Table 16.2 provides a comparison of housing areas in Austin, Texas, based on selected environmental criteria. Most of these features may be assessed from remote sensing imagery under normal viewing conditions. We may appraise housing quality, study certain socioeconomic aspects of a neighbourhood, and estimate the level and distribution of family income on the basis of such a range of urban characteristics.

Perhaps the most important point of all is that the remote sensing view portrays the city as a single system, and the area under special scrutiny as an integral part of it.

16.2.3 Industry
Three aspects of the industrial components of urban areas may be elucidated usefully by aircraft or satellite imagery. These are:
(a) The present location of industry, and different types of industrial land.
(b) The spread of atmospheric and water pollutants away from an industrial complex, and
(c) The opportunities that exist for the establishment of new industries for the re-
    development of existing land under industrial use.

To some extent we have already covered the first of these aspects, since the siting of different types of industrial activity must be embraced by any comprehensive scheme for urban land use classification. At the larger scales of imagery it may be possible to distinguish the component parts of industrial complexes, whether these are single-ownership plants or estates which contain a number of independent factories. However, the potential contribution of remote sensing in this area is comparatively small since new industrial development and the leasing and ownership of industrial premises is subject to closer control than most other aspects of urban land use in the advanced, industrialised nations.

The spread of environmental pollution is much more difficult to monitor by conventional means. A wide range of remote sensing techniques has been tested in the search for efficient methods to measure the concentration and spread of pollution. Examples include:
(a) Visible waveband photography. This has been used for such studies as the behaviour of chimney plumes and the appearance and intensification of urban, grassland and forest fires.

(b) Infrared imagery. This has been used as the basis for studies as diverse as the monitoring of volcanic eruptions, fire detection, and the delineation of warm effluents in rivers. lakes or the sea.

(c) Microwave techniques. These are being used to map oil spills in coastal and deeper waters.

(d) Multispectral processing techniques. These can be used to assess water quality through the different levels of light penetration in selected regions of the thermal emission spectrum.

In general it has been suggested that remote sensing of such aspects of environmental quality in and around urban centres and elsewhere may have two significant contributions to make:

(a) Through the provision of a more complete spatial inventory of areas of atmospheric and water pollution to a single standard throughout a nation, or indeed, the world.

(b) Through the determination of regional, national and global burdens of pollution.

Obviously such broad developments can be contemplated only with satellites, not aircraft, as the sensor platforms. It is worth noting that early experience with ERTS-1 produced encouraging results:

(a) Air pollution. It has been demonstrated that satellite remote sensing can detect particulates emanating from both point (e.g. industrial) and mobile (e.g. aircraft) sources.

(b) Water pollution. It has been possible to map large-scale patterns of turbidity in rivers and oceans using the types and resolutions of data supplied by the ERTS-1 satellite. Water pollution from domestic, municipal and industrial sources has been observed and differentiated.

(c) Land pollution. Large scale landscape problems can be identified, including strip mines, tailing piles and dereliction. However, it is in this area that limitations imposed by the spatial resolution of the ERTS-1 system are most apparent; for example, areas of solid waste disposal are usually quite small.

Turning lastly to the social needs of redeveloping old industrial land it is clear that we must deal with a scale of imagery much better than that achieved by any environmental satellite yet flown, and finer even than much obtained from past aircraft missions. Much old industrial land is rather poor in quality, often as a direct result of its exploitation for industrial purposes. In many cases it is not sufficient just to know the extent of such areas. The precise nature and quality of the land may be equally vital in influencing the new use to which it might be put.

In the United Kingdom, for example, studies of derelict industrial land have been made in the West Riding of Yorkshire (see Table 16.3), using aerial photography at a scale of 1: 10 500. Field checks revealed that a very high degree of accuracy was achieved in the identification of different types of derelict areas within the study area, which measured 200 km². The detailed stereoscopic examination of the prints took

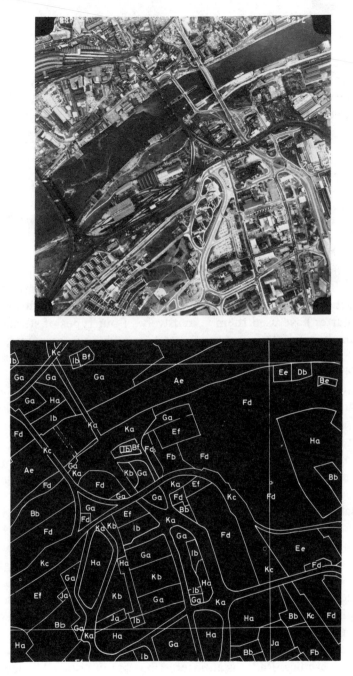

Plate 16.2
An example of aerial photo-interpretation of an urban area: the centre of Newcastle astride the River Tyne. (Courtesy, Fairey Surveys).

approximately 15 man hours. In contrast 15 man weeks would probably have been required for a field survey to collect and map the same information. There is much scope for such studies in the industrialized world, especially where land reclamation is proposed.

16.3 Urban arteries

16.3.1 Transportation networks

Aerial photography has been used successfully for many years in the field of transport studies. In the U.S.A., for example, a survey of major highway organization has revealed that nearly three-quarters of them have made use of aerial surveys in highway planning, although only one-quarter have used them 'extensively'. Four types of studies have been undertaken, namely road planning, traffic studies, parking assessments, and highway inspection.

(a) Road planning. Photogrammetric analyses of air photographs have provided engineers with both qualitative and quantitative data, especially in the early stages of highway routing and design. In some countries sources of road-building materials have been located from the air.

(b) Traffic studies. Aerial photography has been used to pinpoint areas and causes of traffic congestion, and to provide information on traffic flow, both for future road design purposes and immediate ameliorative action. Closed circuit television is used on some urban motorways, for example the Chiswick Flyover in West London, to assist speedy breakdown and recovery operations and help maintain free movement of traffic.

(c) Parking assessments. Aerial photography can reveal where the heaviest concentrations of automobiles tend to build up, and where additional parking spaces might be most beneficial.

(d) Highway inspection. Air surveys are a quick and convenient means of assessing the states of road surfaces which may be in need of repair. In winter the successfulness of snow removal may be adjusted in a similar way.

More recently research has been directed towards other remote sensing techniques for use in transporation studies. Radar is an interesting case in point. All electromagnetic waves are polarized. That is to say, once propagated, they continue to move at a given angle measured against a standard plane of reference unless some outside force or object changes that angle. When radar waves are transmitted horizontally or vertically and are received at the same angle or polarization, they are termed 'like-polarized'. If they are received at a different angle, they are said to be 'cross-polarized'.

Radar is a useful system for delimiting and tracing man-made features in the rural

Table 16.3
An air photo interpretation scheme for the identification and mapping of derelict land in the U.K. using panchromatic photographs. (Source: Bush and Collins, 1974).

No. in air-photo key	Code in derelict land key	Brief description of derelict item
1r	A1ia	Ridge tip
1f	A1ia	Low flat tip
1c	A1ia	Conical coal tip
2	A1ib	Coal dump
3	A1if	Degraded land above ground level associated with coal mining
4	A1ig	Degraded land above ground level peripheral to a coal mine
5	A1ij	Coal sludge above ground level
6	A2ia	A tip of domestic refuse
7	B1ic	Open-cast coal workings
8	B1ih	Open-cast coal workings not yet 'excavations'
9	B1iic	A dry brick clay quarry
10	B1iid	A wet brick clay quarry
11	B1viic	A dry sand and gravel excavation
12	B1viid	A wet sand and gravel excavation
13	B1viie	Degraded land below ground level, resulting from sand and gravel workings, but partially restored
14	B1viih	Sand and gravel workings not yet 'pits'
15	C1ib	Coal dump site at ground level
16	C1ie	Degraded land at ground level, resulting from coal workings
17	C1if	Degraded land at ground level, associated with coal mining
18	C1ig	Degraded land at ground level, peripheral to a coal mine
19	C1ij	Coal sludge at ground level
20	C1viif	Degraded land at ground level, associated with sand and gravel workings
21	C1viij	Sand and gravel sludge at ground level
22	C3if	Degraded land at ground level, associated with a brickworks
23	C3vj	Power station waste at ground level
23a	B1i/3v	Power station waste used to fill a coal excavation
23b	B1vii/3v	Power station waste used to fill a sand and gravel excavation
24	C3vij	Sewage sludge at ground level
25	C4iiif	Railway dereliction at ground level
26	D3i	A disused brickworks
27	l1i	A coal mine
28	l1vii	A sand and gravel works
29	l3i	A brickworks
30	l3v	A power station
31	l3vi	A sewage works

environment because characteristics like flatness, sharpness of corners, and material composition combine to produce stronger returns from cultural than natural features in radar images. In general, cross-polarized signals produce grainier images than the like-polarized, but under certain circumstances some objects in the rural environment are revealed more clearly through them. For example, detecting and tracing communication nets is performed most easily, completely and accurately using cross-polarized imagery when the net traverses the flight path. On like-polarized imagery the best results are obtained when its components are parallel to the direction of flight. Therefore the most efficient system seems to be one in which both like-and cross-polarized components can be assessed. The most useful and obvious applications of such a system are in topographic mapping in developing countries. It is possible that, with suitable improvements and refinements, urban road systems may be investigated by such means, with special reference to their surface materials.

Although the data from Landsat and Skylab are generally rather coarse for transportation studies, the applicability of spacecraft observing systems to transportation geography and linkage analysis has been carefully assessed. In 1965 a broad-based conference held at the NASA Manned Spacecraft Center at Houston, Texas, called for data relating to three broad groups of transportation facts:

(a) Network information. Ideally this necessitates infrared and colour photographic systems, chemical devices, and multispectral sensors, capable of providing data with resolutions of between one and ten metres. From such data maps of the physical linkages and terminal facilities in a network could be compiled. Pattern analyses by computer processes could be carried out once the information had been transformed into matrix and vector form.

(b) Flow phenomena and associated problems. Here infrared and panchromatic photography and radar could be used, giving an optimal resolution of one-half metre. Matters commensurate to investigation would include origin-destination patterns and daily 'tidal' flows of traffic as a whole. These would have useful applications in road design and traffic control.

(c) Relationships between transporation and land use. These necessitate infrared, colour and panchromatic photography, and chemical sensing devices particularly sensitive to phosphorus and nitrogen. The optimal resolution would be one meter. Matters such as the relationships between land-use intensity and distance between road and rail links, land-use capability, and the positions of areas within their larger economic regions would be amenable to study.

Clearly the ranges and resolutions of data generally available to the geographic community are not yet adequate for the achievement of such aims. Nevertheless the list indicates something of the potential of the satellite in transporation studies — as well as some of the hopes and aspirations of would-be data users.

16.3.2 Public services

The use of aerial photographs in civil engineering projects is more than a future visionary dream. For example, for many years aerial photographs have been used successfully in the development of water distribution services, since they provide quick indications of the number of premises to be served, the distances involved, and the types of land use in the service area. Sewage collection is also efficiently planned thereby. Remote sensing provides perhaps the most practical and economic method whereby gradients can be assessed, the layout of pipelines planned, and pumping stations and force mains appropriately sited. Photogrammetric surveys are always conducted in the U.S.A. for the laying of long-distance pipelines and high voltage transmission lines. Microwave transmission towers can be sited best by such means, and their heights selected to meet the necessary requirements for line-of-sight arrangement. In the United Kingdom the Central Electricity Generating Board regularly checks its major power-lines by helicopter-born infrared line-scan (see pp. 62—63). Lastly, remote sensing data are often useful in airport planning and design.

References

Bush, P.W. and Collins, W.G. (1974), 'The application of aerial photography to surveys of derelict land in the United Kingdom', in *Environmental Remote Sensing; applications and achievements,* Barrett, E.C. and Curtis, L.F. (eds.), Edward Arnold, London, pp. 167-183.

Davies, S., Tuyahov, A. and Holz, R.K. (1973), 'Use of remote sensing to determine urban poverty neighbourhoods', *Landscape,* pp. 72-81.

Holz, R.K., Huff, D.L. and Mayfield, R.C. (1969), 'Urban spatial structure based on remote sensing imagery', *Proceedings of the Sixth International Symposium on Remote Sensing of the Environment,* University of Wisconsin, Ann Arbor, Michigan, pp. 819-830.

Horton, F. (1974), 'Remote sensing techniques and urban data acquisition', in *Remote Sensing: techniques for environmental analysis,* Estes, J.E. and Senger, L.W. (eds.), Hamilton Publishing Co., Santa Barbara, pp. 243-276.

17 Problems and prospects

17.1 The limitation of costs on satellite remote sensing studies

The Space Age can be said to have begun on 4 October 1957 when Sputnik-1, the Russian automatic satellite circled the Earth sending back locational bleep signals. Since then there has been a tremendous advance in space technology, mainly as a result of work in the United States and the Soviet Union. Thus the Russian programme provided the first man in space (Gagarin) and the first space walker (Leonov) whilst the American programme culminated in the exciting Apollo moon programme.

All the major advances in space technology have been either Russian or American, though it should be noted that several satellites have been sent up by other nations, chiefly on American and Russian launchers. There has also been the development of major programmes of space research by the European Space Research Organisation (now ESA)*. Nevertheless to the uncommitted onlooker it has seemed to be something of a space race with political undertones, and prestige as a reward. It has also become clear that the programmes were very expensive indeed. Thus the question is raised as to whether such massive expenditure on space research is beneficial to mankind when investment in other forms of industry is thereby limited. Furthermore the world needs for programmes of education, social and medical welfare cry out for investment of available funds. Competition in space may, therefore, be seen by some as wasteful. It could be argued that the needs of the world community would be better served if the space nations collaborated and pursued complementary research activities. Furthermore, even when the desirability of space research is accepted the expenditure is seen by some as being unacceptably high. The most urgent need, therefore, seems to be to reduce the cost of space transportation (cost per kg to low Earth orbit).

The next decade will be an important one in space endeavour because it is during this period that attempts will be made to transform the space frontier of the 1970's into familiar territory, easily accessible to man in the 1980's. In order to achieve this it will be necessary to reduce the cost of transporation by as much as a whole order of magnitude. Important savings can be achieved (perhaps cutting costs by two thirds) if space vehicles can be reused. Further savings could be made if running maintenance were

---

* The European Space Research Organisation and the European Launcher Development Organisation (ELDO) were disbanded in 1975 and a new organisation named the European Space Agency (ESA) was formed.

carried out in orbit. With such objectives and possible means of achieving them in mind present trends in space programmes can be reviewed. First, one may note that the American Post-Apollo programme incorporates the Space Shuttle as a key element for the next decade. This is a re-usuable transportation system which consists of a vehicle that has features in common with an aircraft, a rocket and a satellite. Its fuselage and wing dimensions are comparable with those of a medium-size twin engined commercial jet aircraft e.g. the DC9. It can be brought back from space and land on a runway. The Space Shuttle is designed for rapid re-conditioning and repeated use with a life of 100 sorties or more. The gross lift off weight is more than 2000 tons and after a certain time the solid-propellant boosters are jettisoned. The fall of the boosters is checked by parachutes and retro-rockets so that they can be recovered from the sea, then refurbished and re-fuelled for re-use (see Chapter 5).

In addition to the Space Shuttle an unmanned propulsive vehicle termed the Space Tug has been designed. The Space Tug would be carried into space by the Space Shuttle. It would then be used to transfer a payload from one orbit to another, rendezvous and dock with satellites and with the Shuttle, and be brought back to Earth by the shuttle for refurbishment and re-use.

At the cost of several flights the shuttle could build 'trains' of tugs. Space payloads could then be given the very high velocities necessary for lunar and planetary exploration e.g. around Mercury, Jupiter and Saturn. It has been ascertained that such missions could be achieved more economically by such means than is at present attainable by expensive Saturn rockets.

The Post-Apollo programme is seen as offering the prospect of the re-usable Space Shuttle placing a whole series of unmanned satellites in orbit. It will also enable man to work in space under technical and economic conditions more favourable than they are today. For example, it is possible to envisage a modular-type (i.e. sectionalised) space station constructed in space from separate elements carried up by the Shuttle and then forming a set of working laboratories. In addition there may be a family of research and application modules (RAM) complementing the activities of the station and working with it in close liaison. At present the programme is scheduled to provide the Space Shuttle by the late 1970's and the modular space station with associated RAMs by 1982-1985. This programme seeks, therefore, to provide for some reduction in the costs of space activity, whilst increasing the flexibility of operations in space.

The development of the Post-Apollo programme has been accompanied by interesting examples of international co-operation in space research. First, one may note the European participation in the Space Shuttle programme which was agreed between ESRO and NASA. An agreement has now been reached as a result of which Europe will build the space laboratory (Spacelab) to be used on a co-operative basis in the Space Shuttle programme. Manufacture of the Spacelab has now begun with VFW-Fokker ERNO as the Spacelab contractors. The current position is that the U.S.A. will be responsible for developing, building and operating the Shuttle, and will

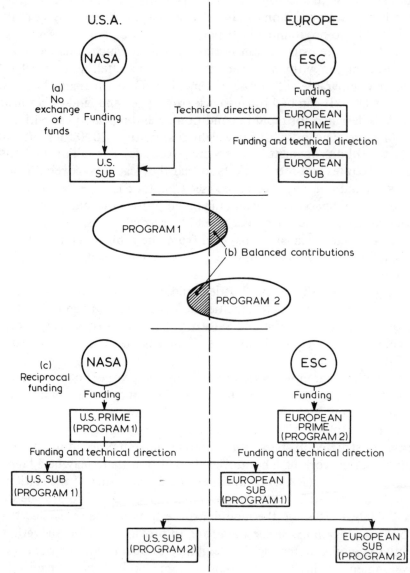

Fig. 17.1

General schemes of management for a co-operative programme between ESRO and NASA.
(Source: ESRO, 1972)

make it available for European use on either a co-operative (non-cost) or cost-reimbursable basis. Europe, on the other hand will be responsible for the design, development and manufacture of Spacelab and associated equipment. The general scheme of management for the co-operative programme is outline in Fig. 17.1

The motivation for participation in the Post-Apollo programme by Europe can be seen to be twofold — the experience and the spin-off. The experience promised by this venture emanates from actual European involvement in an advanced manned space programme with all its attendant status, prestige and challenge. In fact the decision by the Americans to internationalize the Space Shuttle programme has given Europe an opportunity to play a part in an advanced technological development which would not otherwise have been possible because of the high overall cost. The spin-off which might be very large would be of both a technological and scientific nature.

Alongside this example of international cooperation between Europe and the U.S.A. there have been parallel developments of proposals for Soviet—American joint activity in space. In particular plans have been made for exercises in docking procedures and rendezvous methods in space. Such collaboration, if brought to full fruition, could lead to collaborative manning of space stations and also mutual support on occasions of space emergencies.

The above summary of the emerging programmes for remote sensing from space illustrates the great potential in future work. However, there remain some severe problems to be overcome before the full benefits of remote sensing technology can be enjoyed by the ordinary man. These problems lie especially in the broad fields of administration and data processing, and at the interface between user and technologist. Of these it may well be that the problems of administration will place the greatest constraints on the applications of remote sensing techniques in the foreeable future.

17.2 Problems of security restrictions on the use of remote sensing techniques for civilian purposes

When one examines the general administration of remote sensing applications at the present time there are several apparent sources of difficulty. Amongst these there is the continuous restraint, placed by security rulings of various kinds, on civilian use of new tools in remote sensing. Many sensors and sensing interpretation systems are classified as secret and are, therefore, available only to military users. In some cases current security classification rulings seriously inhibit the exploitation of new observation methods for peaceful purposes. The inhibiting effects of security ratings may be summarized as follows:

(a) They restrict the availability of existing data and equipment for present programmes.
(b) They limit the education of scientific and engineering personnel who could use new tools in the future for civilian applications.
(c) They hinder the formulation and development of new research projects in natural science and fruitful interdisciplinary efforts.
(d) They interpose a barrier between large numbers of scientists knowledgeable in

fields where interpretation of sensor output can be useful to both advances in natural science as well as to military reconnaissance applications.

(e) They retard the free flow of technical information within the scientific community encouraging duplication of effort and the reporting of un-criticized findings.

(f) They restrict the development of civilian markets so that system and component costs remain high.

Clearly, this problem is one of balancing potential gains to civilian programmes against potential losses in national defence. The related judgements are always difficult to make but it seems clear that the classification of remote sensing devices as secret should be brought under constant review. Likewise periodic reviews of the rules of classification should be made, as has been done by the Directorate for Classification Management of the Department of Defense, U.S.A.

Interwoven with this problem of the security wraps placed on remote sensing systems there is the other administrative problem of whether 'open sky' observations can be made or not. Some countries are unwilling to allow observations to be made from overflying aircraft and satellite platforms. It is, indeed, praiseworthy that NASA has made weather satellite and Landsat imagery available to Earth scientists with a minimum of constraints. One could wish that more of the imagery obtained by the Soviet Union were easily available to environmental scientists.

17.3 Problems of handling large quantities of remote sensing data

Additional problems arise in the field of data processing as a result of the vast quantities of data that can be gathered by modern remote sensing systems (Fig. 17.2). The Landsat multispectral scanner, for example, produces 15 megabits s^{-1} but new systems will generate at least 10^4 megabits s^{-1} and such high data rates cannot be utilised to the full even by the total global user community. In fact some authorities consider that an operational Earth-observation system, similar to a weather forecast system, should aim at average data streams far below 5 megabits s^{-1}. As a result there is a need for studies of data compression and the extent to which proposed data might be redundant. This requires study of the information content of images, and the establishment of methods of user-oriented redundancy reduction. It will be especially difficult to define the low-redundancy thresholds in terms of physical quantities like spatial limit, frequencies, dynamic range etc.

Another approach to the problem of the quantity of data is to adopt some means of generalizing data subsequent to its collection. For example, some workers have adopted unit test areas (UTAs) of 50 km square for generalisation purposes. The data for each UTA is reduced to a single number for each spectral band. Preliminary results suggest that for most environmental applications this would be too drastic a reduction in the information content of the data.

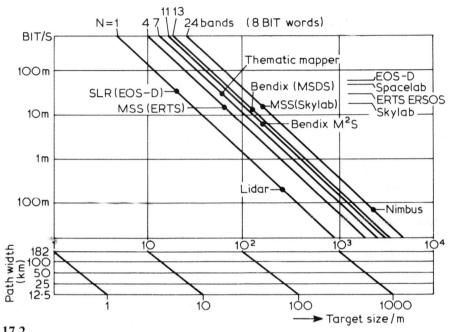

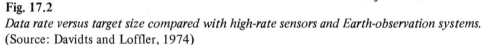

Fig. 17.2

Data rate versus target size compared with high-rate sensors and Earth-observation systems.
(Source: Davidts and Loffler, 1974)

17.4 The interface between technologist and user

Another major problem area is that of the interface between the remote sensing technologists and the users of remote sensing data. Defining the term 'user' is somewhat difficult in practice because there are many users with different objectives. Broadly speaking, however, we may say that there are two main groups of users, the academic and the operational. In the academic category one can include Earth scientists in universities and other research institutes. On the other hand, operational users are typified by agricultural departments, meteorological services, geological surveys and mineral companies. Each of these groups of users finds difficulty to a greater or lesser extent in:

(a) Comprehending the nature and significance of remote sensing data obtained by non-photographic sensors.
(b) Defining his own requirements in terms of the engineering and physical properties of sensors used in remote sensing.

It is also apparent that the space technologist does not always fully comprehend the dynamics or properties of the environmental surface being sensed. Furthermore as increasing use is made of non-visible waveband imagery rather than conventional photography the problems of communication between the remote sensing technologist and the general user become more severe. To the uninstructed person an image obtained by infrared or microwave systems looks like a photograph. Its record is, however, an entirely different information array from that obtained by photography. For example in microwave studies the measured response (image) obtained from a land surface is affected by complex factors which can be grouped under two headings:

(a) Characteristics of the microwave system
- polarization direction
- observation angle
- frequency of bands used.

(b) Characteristics of the surface sensed
- electrical and thermal properties
- surface roughness and size
- temperature and its distribution.

In these circumstances it is clearly desirable that there should be a flow of information from the physicist/technologist to the Earth science user. However, it is equally true that many physicist/technologist personnel have very limited conceptual and mathematical models of the Earth phenomena being sensed. This has sometimes resulted in exaggerated claims being made for the usefulness of remote sensing systems. For instance some early writers on air photography and soil studies led potential users to think that soil mapping could be achieved easily by air photo interpretation. Such statements were based on a lack of knowledge of soil classification and how classificatory techniques affected the objectives of air-photo interpretation. Lack of environmental training has also resulted in some researchers collecting second rate data concerning the land surface conditions (e.g. in respect of soil moisture – see p. 240) for comparison with highly accurate remote sensing data.

It is highly desirable that the teams working in remote sensing should be of a multi-disciplinary character from the outset. If eventual users were involved at the beginning they could help to solve many of the technical problems such as the type of output required. Such participation would also allow the user to make an objective evaluation of how to make best use of the information. The need for such involvement by the users is underlined by this comment made in respect of crop inventory studies during a symposium on results from ERTS-1: 'from the standpoint of applications, almost every investigation lacked complete definitions of techniques and procedures which would allow a quasi-operational project to be undertaken'.

One feels that remote sensing systems have grown largely without regard to the needs of the community which may use them. Perhaps this is understandable at the early stages of a massive research programme with so many ramifications. It becomes less acceptable when a system which could bring great benefits to man has failed to achieve its full potential. There surely comes a point at which potential users should be given an opportunity to mould the system towards their own ends and also have an opportunity to adapt their administrative and social programmes to allow use to be made of the remote sensing data collected.

Some potential user agencies have been criticized in the past because they do not appear to recognise and understand the uses of remote sensing data. This criticism will doubtless continue to be voiced so long as the planning and execution of remote sensing studies does not involve the users at an early stage. It should be recognized that fruitful use of remote sensing data depends not only on successful scientific advance but also on budgetary and administrative considerations within the user organisations. This often shows itself when studies of cost effectiveness are made. One frequently hears the question 'Would it not be cheaper and easier to send out a man in an automobile to collect information ?' put at a late stage in an investigation rather than at its beginning. Quite often the answer to the question is 'no', but sometimes the economics of the case are less clear and then the accusation of 'technology chasing users' seems apt. It seems fair to add, however, to those who make such comments that innovation can scarcely ever be proved to be cost effective. Thus the first move towards progressive technology is almost certainly more expensive than status quo.

Clearly these issues are matters of judgement and the decisions are often difficult to make. In these circumstances it is important that facilities should be available to educate the decision makers and inform the general public about the possibilities and limitations of remote sensing techniques. The educational facilities are rather unevenly spread at the present time. University students are more likely to have been instructed in some aspects of remote sensing studies than school leavers. Geography departments are likely to include some formal studies of remote sensing for environmental monitoring purposes but teaching is likely to be fragmentary in other disciplines.

At the present time symposia and conferences tend to bring together groups of specialists in remote sensing. These are greatly needed and fulfil a useful purpose in that they are becoming increasingly multi-disciplinary. Also the journals devoted to remote sensing studies are embracing a wide range of contributions from different disciplines. This is true of the long established journals such as *Photogrammetric Engineering* (U.S.A.) and *The Photogrammetric Record* (U.K.) as well as newly constituted journals such as *Remote Sensing of the Environment* (U.S.A.).

The increased interest in remote sensing studies has resulted in the formation of new societies such as the Remote Sensing Society in Britain and embryonic groups in Europe and Australia. These, together with existing societies such as the Photogrammetric Society (U.K.) or American Society for Photogrammetry (U.S.A.), provide the organizational frameworks which can handle enquiries and provide sources of information.

Information flow has also been assisted by the compilation of directories of activities in remote sensing by the Department of Industry (D.O.I), U.K. and the European Space Agency (ESA) and the widespread circulation of study reports by the National Aeronautics and Space Administration (NASA). Indeed much of the stimulus has come from national and international institutions and government-funded multidisciplinary studies such as those carried out by the Natural Envrionment Research Council in 1971.

This book has sought to provide a base from which the reader can proceed towards greater understanding of the part that remote sensing techniques can play in our lives. If it has opened some windows on the opportunities and pitfalls that lie ahead in the use of remote sensing for earth resource development, the authors will be well pleased. No doubt the ultimate goal will be a computerized satellite system as shown in Fig. 17.3 — but if that is ever achieved one may wonder whether man is fully equipped to use the leisure and knowledge which such automated control might provide for him.

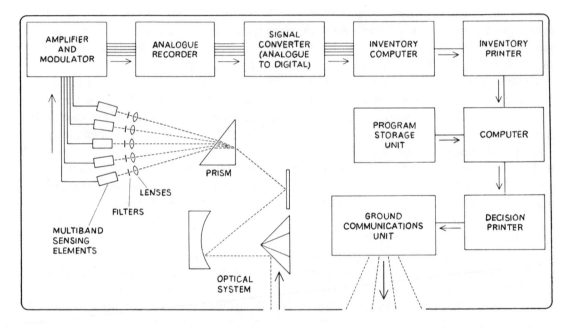

Fig. 17.3

Computerized satellite system for the future. It would sense resources in several wave bands, automatically identify them, weigh them against previously programmed data on the cost effectiveness of various management possibilities, and send to the ground a decision on what should be done. It also could be used to monitor developing situations, such as a forest fire, suggesting how ground crews might fight it, and to perform automatically such tasks as turning irrigation valves on and off as required. (Source: Holz, 1973)

References

Curtis, L.F. (1963), 'Soil classification and photo interpretation', *International Archives of Photo-grammetry,* **14**, 153.

Curtis, L.F. and Meyer, A.E.S. (1974), *Remote Sensing Evaluation Flights, 1971,* Thetford Area, Natural Environment Research Council Publication Series C. No. 12, pp. 1-50.

Davidts, D. and Loffler, A. (1974), 'Reduction of information redundancy in ERTS-1 and EREP Data', *Proceedings Frascati Symposium on European Earth-Resources Satellite Experiments,* European Space Research Organisation, pp. 81-91.

ESRO (1972), *Report on European Participation in the Post-Apollo Programme,* European Space Conference, WG/COOP/US (72) 2.

Holz, R.K. (ed.), (1973), *The Surveillant Science,* Houghton Mifflin, Boston.

Lusignan, B. and Kiely, J. (1970), *Global Weather Prediction; the coming revolution,* Holt, Rinehart and Winston, New York.

Simmons, N. (1974), 'The Remote Sensing Society', *Estratto Degli Atti Ufficiali Del XIV Convegno Internatzionale Technico-Scientifico Sullo Spazio,* Rome, III-6, 209-215.

Suits, G.H. (1966), 'Declassification of infrared devices', *Photogrammetric Engineering,* **32**, 988-992.

Index